Teaching and Learning Technology

Teaching and Learning Technology

Edited by
Robert McCormick, Patricia Murphy
and Michael Harrison
at the Open University

Addison-Wesley Publishing Company

Wokingham, England • Reading, Massachusetts • Menlo Park, California
New York • Don Mills, Ontario • Amsterdam • Bonn • Sydney • Singapore
Tokyo • Madrid • San Juan • Milan • Paris • Mexico City • Seoul • Taipei

in association with

The Open University

Many of the designations used by manufacturers and sellers to distinguish their products are claimed as trademarks. Addison-Wesley has made every attempt to supply trademark information about manufacturers and their products mentioned in this book.

Cover designed by Designers & Partners of Oxford
printed by The Riverside Printing Co. (Reading) Ltd.
Typeset by Columns Design and Production Services Ltd, Reading, UK
Printed in Great Britain by T.J. Press, Padstow, Cornwall.

First printed 1992.

This reader and its companion volume *Teaching and Learning Technology*, are part of the course material for an Open University M.A. in Education module, E823 *Technology Education*. Open University students have access to other material as part of this module and hence the contents of this reader are necessarily a limited selection.

Anyone wishing to obtain more details about the module and the M.A. programme should contact The Central Enquiries Office, The Open University, Walton Hall, Milton Keynes MK7 6AA.

The views expressed in the articles in this collection are not necessarily those of the team responsible for developing the module, nor of the Open University.

British Library Cataloguing in Publication Data
A catalogue record for this book is available from the British Library.

ISBN 0–201–63169–5

Acknowledgements

The publisher has made every attempt to obtain permission to reproduce material in this book. Copyright holders of material which has not been acknowledged should contact the publisher. Grateful acknowledgement is made to the following sources for permission to reproduce material in this reader:

Modified version of article published in Hacker, M., Gordon A. and De Vries, M. (eds) *Integrated Advanced Technology into Technology Education*. Berlin: Springer-Verlag/NATO Scientific Affairs Divn. Scottish Consultative Council on the Curriculum (SCCC) (1989) Curriculum Design for the Secondary Stages. 1st revised edn. Edinburgh: SCCC. National Science Foundation (1983) *Educating Americans for the 21st Century*; National Science Foundation (1983) *Educating Americans for the 21st Century – Source Material*. Washington DC: NSF. Johnson, J.R. (1989) *Technology: Report of the Project 2061 Phase 1 Technology Panel*. Washington DC: American Association for the Advancement of Science (AAAS). Savage, E. and Sterry, L. (eds) (1990) *A Conceptual Framework for Technology Education*. Reston VA: International Technology Education Association. Black, P. and Harrison, G. (1985) *In Place of Confusion: Technology and Science in the School Curriculum*. Nuffield–Chelsea Curriculum Trust/National Centre for School Technology, Trent Polytechnic. Woolnough, B. (ed.) (1991) *Practical Science*. Milton Keynes: The Open University Press. APU (eds) (1991) *The Assessment of Performance in Design and Technology*. London: School Examinations and Assessment Council. McConnell, M.C. (1982) Teaching about science; 9, 1–32; *Studies in Science Education*, 1991, 19 63–75 (C) Studies in Science Education Ltd. Rouse, J. (1987) *Knowledge and power*. Used by permission of the publisher, Cornell University Press. Rogoff, B. and Lave, J. (eds) (1984) *Everyday cognition*. Cambridge MA: Harvard University Press. Wilson, S., Shulman, L. and Richert, A.E. (1989) 150 different ways. In J. Calderhead (ed.) *Exploring teachers' thinking*. London: Cassell. Rogoff, B. (1990) *Apprenticeship in thinking*. Oxford University Press. Williams, D. (1985)

Understanding people's understanding of energy use in buildings. In Stapford, B. (ed.) *Consumers, Buildings and Energy*. Birmingham University. *American Psychologist*, 1984, 39 (2), 93–104: American Psychology Association. Brown, C. 'Girls, Boys and Technology' published originally in *The School Science Review* (1990), 71, No. 257, pp.33–40. Siegler, R.S. and Richards, D.D. (1982) The development of intelligence. In Sternberg, R.J. (ed.) *Handbook of human intelligence*. Cambridge University Press. Minsky, M. and Papert, S. (1974) *Artificial Intelligence*. Eugene: Oregon State System of Higher Education. *International Journal of Science Education*, 1989, 11, pp.481–90. *The School Science Review* (1990), 71, No. 257, pp.33–40. *International Journal of Technology and Design Education* (1990) 1 (1), 14–20. Cowie, H. and Rudduck, J. *Cooperative Group Work: an overview*. London: BP Educational Service. Jacques, D. (1984) *Learning in Groups*. London: Croom Helm. Fensham, P. (ed.) (1988) *Developments and Dilemmas in Science Education*; Harding, J. (1986) *Perspectives of gender and science*; Jamieson, I., Miller, A. and Watt, A.G. (1988) *Mirrors of Work: Work simulations in schools*. London: Falmer Press. Fleming, R. (1986) Adolescent reasoning in socio-scientific issues part 1. *Journal of Research in Science Teaching*, 23 (8), 677–88 John Wiley & Son Ltd. Shapcott, D. and Wright, V. (1976) The Christmas Cracker Factory. *Careers Adviser*, 5 (4). DES/WO (1987) *The National Curriculum 5–16: a consultation document*. London: Department of Education and Science/Welsh Office. Reproduced with the permission of the Controller of Her Majesty's Stationery Office.

Introduction:
In Search of Technology Education

R. McCormick, P. Murphy and M. Harrison

Technology education for all pupils is a relative newcomer to the curriculum. Many of the battles that were fought long ago over the introduction of science into the curriculum are being, or have recently been, fought over technology. Finding a selection of articles that represents the range of views, models and activities that currently comprise technology education has been a challenging task, not least because technology education is not yet as well documented as some longer-established areas of the curriculum.

The companion volume to this reader, *Technology for Technology Education*, sets out to give a broad view of the nature of technological activity in the world. We believe that a prerequisite for successful teaching of technology is that teachers must have thought through their own understanding of technology in the world, and come to some conclusions for themselves about what it is and how it should be taught. But the last decade has seen rapid developments in technology education in all parts of the UK, and this reader sets out to explore how technology education is currently represented in the curriculum, and also what is known about how technology may be learnt and assessed.

Aspects of technology education have long been present in a range of school subjects, some of which include 'technology' in their title, and some of which do not. So the articles in this reader have been drawn from sources that represent this wide contributory base. We believe that they provide a valuable resource for anyone who wishes seriously to consider the pedagogic implications of implementing a broad technological curriculum for all pupils.

The articles are grouped in five parts. Part 1 begins by setting a context in time and place for current initiatives in technology education. After looking in more detail at the proposals and models now applying in the various parts of the UK, the section concludes with articles about the interestingly different interpretations in the Netherlands and the USA. Part 2 contains four articles that address issues relating to the nature of what has become known as technological capability. Starting with an extract from Black and Harrison's original thoughts on capability, the part continues with articles that pose questions about the practical action that is a central feature of capability. The four articles in Part 3 explore the background to the way children learn technology, ranging through investigations into children's work in science and maths, practical work, and work with computers. Part 4 develops the theme of children's learning into several areas of direct interest to the teaching of technology. Part 5 concludes the reader by considering a range of teaching methods and approaches appropriate to developing aspects of technological capability. It also addresses issues of how teachers respond when faced with a requirement to begin teaching something new.

The articles have been selected on the basis that they each have something important to communicate about technology education. Some are easier to read than others: this is because some aspects of technology education are easier to communicate and discuss than others! We believe that all the issues addressed are of importance at a time when technology education for all is becoming a reality.

Contents

PART 1

General Ideas on Technology Education

1.1

The Evolution of Current Practice in Technology Education

R. McCormick

[This article was originally presented as a very much longer unpublished conference paper. In this version of the shorter published paper we have removed most of the references because of the limitations on space, and rearranged it to suit its use in this Reader. (You should refer to the published version if you want the justification of the author's argument.)]

■ Introduction

Technology education has become recognized as a distinct area of education, and there is increasing international convergence of what constitutes technology education. While there is convergence at the general level of policy documents, there remain differences of emphasis in practice. These differences are illuminated by the way in which the various *traditions* of teaching technology and *reasons* for teaching technology have combined in particular countries. I shall argue in this paper that:

(1) We must learn from the various *traditions* because they encapsulate strongly held views and years of experience that will remain even after we have an established area of technology education.

(2) We need to be aware of the various *reasons* for teaching technology, because they have provided support for the traditions and provided rationales and agendas that come through in their practice.

I will conclude by reiterating the plea made by McCulloch, Jenkins and Layton (1985) for a sensitive view of curriculum change, avoiding the naïvety of some who have tried to introduce technology into schools unaware of the forces exerted by the various interest groups that operate at both national and school level. In making this case I will build on my earlier work (McCormick, 1990), and that of Peter Medway (1989). Together the various traditions of teaching technology, and their associated reasons, encapsulate much of what is coming to be accepted as technology education. The problem is that they have not yet come together sufficiently, and this article attempts to help this process by clarifying the nature of, and the relationship among, the traditions.

■   The traditions of technology education

□   Interest groups

The various traditions forming interest groups, according to curriculum historians such as Goodson (1988), determine the development of curriculum subjects. In this section I want to examine what each of the traditions contributes to technology education practice, relating it where appropriate to the conflicts and debates among the groups within each tradition, and between the traditions themselves. These conflicts and debates at national level will shape technology education, and their adherents at school level will, along with the interaction of the practicalities of school life, reflect this shaping.

The five traditions I have chosen to consider are: craft; design and art; science; science, technology and society; and those who are concerned for industry and productive work. They are not an exhaustive selection, nor are they of equal concern in all countries.

□   Craft

This tradition is not a single one and differs across countries and time. Three strands are identifiable. One strand is (trade) craft stemming from the 19th and early 20th century manual training, and emphasizes exercises to develop skills with tools. Another strand of craft developed from Swedish *sloyd*, that in some forms became aligned with art. The third strand is the development of manual training into industrial arts, with a concern for reflecting the developments in industry. There have been some interesting international exchanges of ideas, though not always with due consideration for the context.

Some advocates of Swedish *sloyd* based their case upon the child-centred approach of progressive education, and were concerned with free creativity, using Dewey's work as justification. Such advocates tended to ignore the social reconstructionist side of Dewey (Skilbeck, 1976) that Luetkemeyer sees as the foundation of industrial arts in the USA (Luetkemeyer, 1985). In contrast to the UK craft tradition, ossified in craft reminiscent of the 19th century, industrial arts in the USA attempted to reflect industrial developments, although that did not continue into the middle of the 20th century.

The difficulty for US industrial arts, UK crafts, or for school vocational education, now, as in the 19th century, is to reflect current industrial practice, which is the main context for most modern technological *activity* (e.g. designing and making). The trade craft tradition has of course developed from wood to metal, from hand tools to machinery, but is unable to continue to develop as technology develops because the basic model of production is flawed, and because there is a limit to what can be done in schools. The craft model of the single worker in control of the whole production process, that Swedish *sloyd* specifically sought to preserve at the advent of industrialization, does not reflect the team work of industry. [See Reader 1, Article 2.3.] The failure of the use in schools of computer-controlled machines, that ignores the context of a modern factory, is exacerbated as flexible manufacturing systems (FMS) become more common in industry. A craft teacher, used to the immediacy of the workshop, sees production games and simulations as theory rather than practice. One response to this is to focus on the process through problem solving and design, an approach which picks up on the progressive education tradition of earlier times and is present in some English primary schools.

But the difficulties that the craft tradition may have should not blind us to the important contribution it makes to how to teach skill development, use materials, and manage individually-based project work. The emphasis of this tradition on finished products does, however, create problems for the development of modelling skills.

☐ **Design and art**

The linking of these two is to some extent arbitrary, but is in part a reflection of the British context within which *design education* developed. There are three main interrelated British interest groups (not all of them homogeneous): those from art and design, design education as part of general education, and Craft, Design and Technology (CDT). The interrelations are not altogether clear to me, as there has been less research on them than on other groups.

In the 19th century, two groups existed within art and design. One, the Society of Art Masters (SAM), was a hierarchical, subject-centred association of art school (male) principals dedicated to presenting drawing as an academic discipline, emphasizing classical draughtmanship and design allied to industrial arts. The other, the Art Teachers Guild (ATG), contrasted strongly with SAM. It was mainly made up of female classroom teachers, and was egalitarian, individual-centred (child-centred) in its interests. Its outlook was to support those who considered the creativity, expression, invention and imagination of childhood as important. This was a by-product of the English Arts and Crafts Movement, with its concern for the spiritual value of craftwork. Industrial art to them suppressed free creativity. In some senses the differences of SAM and ATG resemble the conflict of the manual trainers and the Swedish *sloyd* of about the same period, although I have no evidence to suggest that there was any connection. The successors to these two organizations combined in 1984 to form the National Society for Education in Art and Design (NSEAD), in a synthesis of their two approaches. Currently NSEAD sees itself as representing art and design interests to government, and in that role commented upon the proposals for the National Curriculum for Design and Technology in England and Wales.

Those seeing design education as part of general education focused initially around the Royal College of Art and Bruce Archer, who was responsible for an enquiry into design in schools. He viewed design as a third area in education, distinct from, but equal to the sciences and humanities. [See Reader 1, Article 1.4.] This area concerns doing and making, and includes such things as technology and fine, performing and useful arts; he represented technology as lying somewhere between science and design. Those who see technology as an independent area would put design and technology on a more equal footing. As an interest group they are represented by the Design Council and, along with other groups, are represented on the Design Education Forum. This Forum collectively impressed upon the working group, drawing up the Design and Technology National Curriculum, the importance of seeing design, not technology, as the overarching curriculum activity. Such views have contributed to an understanding of design. In particular, they offer a sophisticated view of design based on what practitioners do and, given their art background, they not surprisingly want to emphasize imagination and intuition. They have also emphasized the importance of visualization, not just as a mode of aesthetic expression but as a way of thinking giving a sophisticated view of modelling.

CDT developed under a number of influences, including that of design education (Penfold, 1988), and the view of design which CDT has developed, epitomized by John Eggleston's work (Eggleston, 1976), has not surprisingly been criticized by the art and design groups. These criticisms include: a mechanical view of design, expressed through various linear and circular staged models that fail to teach pupils the process, and do not reflect how designers think; a view of craft that emphasizes an outdated woodwork and metalwork view of making, rather than one which develops new ideas and designs; and the problem of modelling I mentioned earlier.

Viewing these strands at school level, primary school teachers, for example, may find that the design education approach fits their progressive ideology, but lacks any appreciation of what design is like to the professional designer. They may align it with their views on self-expression in art, without the synthesis arrived at by NSEAD.

☐ **Science**

This tradition is only now trying to come to terms with providing 'science for all', rather than for specialists. Having started with 'applied' beginnings in the

19th century, science educators in England went through a post-war phase of emphasizing 'pure science', and were reluctant to support technology education. Now the Association for Science Education (ASE) recognizes the importance of technology education, as different from science education, and the role of science teachers in it. This development is mirrored internationally, though science education interests still seem to dominate, and this is replicated by the mainstream science interests of both the British and American Associations for the Advancement of Science.

The concern to make science relevant has increased the interest in studying science in meaningful contexts, technology being one of these. Problem solving, therefore, becomes important. Science brings to technology education: scientific investigation and experimentation methods including ideas of a 'fair test'; a knowledge and conceptual understanding essential to a technologist; the development of new science and technology areas such as biotechnology, that CDT and home economics teachers, for example, are often ill-equipped to handle. Science educators have also built up an impressive empirical research literature from which we can learn and build upon to understand the unique problems in learning technology.

☐ **Science, technology and society (STS)**

This tradition is one response by those in science education who want to teach science in context, to avoid the irrelevance of 'pure science', and by implication to replace 'science for all' in schools by STS. Support among science teachers no doubt accounts for early STS courses not distinguishing between science and technology, but now this situation is improving. The STS tradition has been responsible for developing value issues in relation to technology, and, in particular, has developed teaching material and methods for discussing controversial issues, role-play activities and games [see Article 5.4], methods more familiar to social studies teachers. This tradition also brings in a different kind of problem solving from that of design-and-make, one which relates to decision-making about, for example, where to locate a power station.

☐ **Industry, productive work and education**

In this subsection I will not adopt a historical perspective because industry and work is less a tradition than a common theme for a variety of groups. Marxist and 'western' industrial interests share some things in common, and here I will try to illustrate the resulting practice. Both take ideological stances, though of quite different kinds. Marxists are concerned to show the creative activity of production, along with an understanding and control over the production process (part of polytechnical education). Western industrialists argue for the importance of industry to wealth creation, and the need to develop business and entrepreneurial skills and attitudes. Thus in England we have education for 'economic and industrial understanding' as a cross-curricular theme. For both groups, two interrelated aspects are important: an understanding of industry/ production and a change of attitude of students to it. I will deal with each of these in turn.

Understanding industry   For technology education the significance of under-standing industry is twofold: to see industry as a manifestation of modern technological activity; to see industry as a part of culture, as a working environment. I have already noted the limitations of craft work as a model for modern technological activity. It may give experience of design-and-make, but it cannot illustrate properly the industrial process. For example, at the heart of the current concern for quality in industry is 'process control', and the methods appropriate to it (e.g. flowcharting and critical path analysis) are every bit as important as design, yet they are difficult to apply to craft-based activities, particularly individual student projects. Although the UK has a variety of support agencies, the tradition of using manufacturing, at least as a knowledge area, is better in the USA. It developed, as I noted earlier, from early 20th century analyses, and is one of the standard areas of technology originating from the Jackson's Mill analysis in 1981 (Hayles and Snyder, 1982). Manufacturing was also singled out in the *Project 2061* technology panel report as one of '*the* technologies' (Johnson, 1989, pp. 16–17).

The second significant feature of industry, as part of culture, is less well used in technology education. Work is a major part of life for most people and they ought to understand that part just as any other aspect of culture. Industries are complex organizations, with a variety of concerns from technical to environmental and economic. In a sense what we need is an STS of the inside of the factory, rather than just looking at its external impact as would be the case with environmental studies. The area of, for example, industrial relations has been more the concern of social studies or careers teachers, rather than technology teachers.

Student attitudes   Both Marxists and western industrialists want students to have positive attitudes to industry, for their different reasons. The experience in communist countries seems to be poor, although little measurement of attitudes to work has taken place. In the west the excellent studies of PATT (Pupils' Attitude Towards Technology), reported in the annual conference reports, do not deal directly with attitudes to industry and have not to my knowledge yet looked at the effects on attitudes of activities that involve industry. Studies of introducing industrial applications into science teaching, to make it more relevant and interesting, are, however, not encouraging.

There are a variety of approaches to teaching about industry, including games, simulations, role play, case studies and visits, and dealing with problems given by industry, but they all have to face the difficulties of providing first-hand experience, and realism. These are the central problems given the nature of modern industry, as I noted earlier with regard to FMS.

☐ **Learning from the traditions**

The above account of the traditions that have contributed to the evolution of technology education is obviously not exhaustive. Teachers of home economics parallel craft teachers in their development from 'cookery and sewing' to food science and technology and textiles. Business studies teachers are now being asked to contribute, especially when industrial and commercial aspects are concerned. Then there are the geography teachers for environmental issues! The potential list is long, and the selection of traditions I have made serves the purpose of showing the importance of traditions. No single one holds the answer to developing technology education, and it is doubtful if any one could take on the guise of all the others. Nevertheless there has to be a sharing of perspectives, always mindful that particular reasons for teaching technology will lead to different kinds of programmes. Although I have been discussing these traditions at the level of national or international groupings, they have their correlates in schools. Most obviously they exist in particular subject and departmental groupings and in the interests, enthusiasms and skills of individuals.

■ Reasons for teaching technology

The reasons for teaching technology emanate from various interest groups, such as employers' organizations, science bodies (Royal Society or British Association for the Advancement of Science – BAAS), engineering interests and teachers' associations, with the driving force to promote their view of technology education. The reasons for teaching technology that are dealt with here are those concerned with: improving the economy (the economic imperative); educational benefits for the personal development of the individual (intrinsic value); creating responsible citizens (citizenship). I will also consider briefly the political ideological reason that derives from Marxism.

☐ **The economic imperative**

This is historically the most important, and can be seen in the arguments for manual training at the end of the 19th century. In England, various commissions considering technical education contained industrialists who advocated manual training in an attempt to stem the decline of industry in the face of competition from the USA and Germany. In 1979 Britain began the Great Debate on education prompted by the perceived failure of schools in the face of a declining economy. This led to the launch of the Technical and Vocational Education Initiative (TVEI). This provided enormous funds and gave a boost to technology education, though of only a limited kind.

In the USA at the end of the 19th century, similar concerns were expressed about the economy, leading to recommendations for manual training to aid the development of American industry. During the 1980s concerns over the state of industry have re-emerged, leading prestigious bodies to produce reports (such as *Project 2061*) and programmes to improve technology and science education. In European countries with vocational or pre-vocational education, similar economic justifications are used.

Although it is a persuasive reason, its perceived support for technology education is complicated by a number of arguments about the relationship of education and economic improvement. I can do no more than list them here:

(1) Vocational versus general education: Is vocational education better for economic development than general education (probably not)? Is the distinction between the two kinds of education valid?

(2) Status of vocational education: What about the inequity of selection for vocational education and elitist views of academic education?

(3) Science versus technical/technological education: How do the various science and engineering organizations and industry view the merits of each?

(4) Do we change students' attitudes to industry or give them general-purpose skills and flexibility for future industrial technology?

The purpose in identifying these arguments is that they complicate the case for technology education, when it draws upon the economic imperative, and are a mixed blessing. In the UK, for example, employers have shown little interest in supporting vocational technical education in schools, and have preferred science education rather than technology, albeit science with applications. They have, however, supported campaigns on improving attitudes to industry (e.g. Industry Year in 1986), and have been more interested in school leavers' personal qualities than in their formal qualifications.

The support in the USA for the approaches to education in the competing economies of the Federal Republic of Germany (as it was) and Japan ignores the particular stance adopted to each of the above arguments in those two countries. Japan has little vocational education, very academic general education, and does no technology in schools. Germany has well developed vocational education, and technology is just being introduced into schools.

Despite such unhelpful evidence, the economic imperative is seductive and it has at times provided the motivation for the funding of education or training initiatives by governments and industry.

☐ **Intrinsic value**

This reason for teaching technology is one which sees educational benefits, by contributing something to the development of an individual that other curriculum areas cannot. In the 19th century, when industrialists were pushing for manual training to improve the economy, teachers and other educationalists were arguing for its moral, physical, motivational and pedagogic benefits to students. This reflected their desire to raise their status by using the kinds of arguments that teachers of other established subjects might employ.

More recently the problem-solving inherent in technology education has been seen as an educational justification, relying on Dewey's ideas on reflective thinking. The ITEA took this stance in the phrase 'humans must act in order to be' (ITEA, 1988, p. 8). A second element in the problem-solving justification is that 'real' problems require a multi-dimensional approach, and counter the fragmentation and artificiality of a subject-based curriculum. This, so the argument goes, is more in tune with the child's mind. This is the view taken by those who support a progressive education ideology, and has been instrumental in encouraging primary school teachers in the UK to take up technology education. It is also the kind of view taken by some who support design education.

There is a further intrinsic justification: that it is necessary to learn technology problem-solving techniques, in the same way as we teach scientific investigation. This seems to be the implicit approach in the USA. Along with this is the concern for 'doing' rather than 'knowing', and 'knowing how', rather than 'knowing that', reflecting the 'tacit knowledge' of the craft tradition. This last concern has led to the idea of 'capability', which in the UK has been supported by a group of influential people under the auspice of the Royal Society of Arts.

All this is a long way from the hand–eye justifications of a hundred years ago, but is nonetheless attributing the same intrinsic value to technology education as was done for handwork in schools.

☐ **Citizenship**

This reason for teaching technology has two major overlapping strands. The first is a concern for the impact of technology and the role of the ordinary citizen in decisions about technology, in particular controlling it. The second strand concerns the need for people to survive in a technological world.

The concern for the impact of technology has its roots in the environmental movement associated with the 'limits to growth' arguments of the 1970s, and also with the 'social responsibility for science' movement. These concerns were taken up in social studies, and in 'science, technology and society' (STS) courses in both the USA and UK, and early ones were anti-technology.

The value-laden nature of technological activity, and the need to reflect this in education, is now accepted in most proposals for technology education, but it is not yet found in some areas of technology teaching. Science, environmental and social studies teachers have historically been more likely to deal with value issues, than craft or technical teachers.

The second strand, of preparing future citizens to survive in a technological society, embraces the first strand, because it wants to prepare young people for all aspects of life that technology will affect. *Understanding* and *using* technology are important for this strand, and technological 'awareness' or 'literacy' are key terms in the UK and USA respectively. Technological literacy may not be fully defined yet, but its definition reflects two groups in the USA: STS supporters represented by The National Association for Science, Technology and Society (NASTS), and technology educators represented by the International Association for Technology Education (ITEA). They are not, however, entirely distinct, as Cheek acknowledged in his introduction to the international technology literacy conference, that acts as the annual meeting of NASTS (Cheek, 1989).

☐   Marxism

This has been a powerful force in the education systems of many countries. Marx was concerned about the impact of large-scale industry in the 19th century and his analysis of the position of the worker reveals some interesting points. He had remarkable foresight in seeing how craft models of work were being replaced when, 20 years later (in the 1880s), industrialists were advocating manual training in schools. To prevent workers being enslaved by machines he argued for mobility and flexibility, to achieve a 'totally developed individual' (Marx, 1976, p. 618), something advocated by present day industrialists. To achieve this he advocated handwork *and* technological education, which later became known as 'polytechnical education'. Thus we have an early recognition of technology education, and one which informs the practice of many of the Eastern European countries with which we will be increasingly involved.

A second element from the Marxist tradition comes from ideas about the role of productive work in defining the nature of human beings, who create the world they inhabit and in so doing are changed by it. This parallels the ITEA view noted earlier (humans must act in order to be). Labour is therefore seen as important as part of education. In addition to this *intellectual* objective of productive work, there is the *ideological* one of wanting to overcome class differences and the distinction between mental and manual work. The argument about the status of technical versus academic education, noted under the economic imperative, has much in common with this latter distinction.

This objective requires changes in attitudes to work, and again there are similarities with western industrialists' concerns, albeit from a different ideological standpoint.

☐   A mixture of reasons

Those who have advanced the case of technology education have done so on the basis of one or more of the above reasons. The use of the reasons may vary over time and across the traditions that go to make up current practice. Thus science education has used economic, citizenship and intrinsic arguments to justify its own place, adopting a colonial stance towards technology. Similarly design education has called upon intrinsic and economic arguments, in the same way as manual training did at the beginning of the century. To some extent this may be opportunism, but also different factions within traditions, and at different times, change the climate of rationales. The economic reason, although the most potent, causes the most difficulties as a basis for justification. For example, the use of the economic imperative to justify a technology programme which emphasizes understanding technology in the world, and associated social issues (as an STS course might), would not satisfy those who sought to improve manufacturing.

■   Curriculum change

By way of a conclusion to this article let me return to something I signalled in the *Introduction*: the need for a sensitive view of curriculum change. This must be one that recognizes that it is not enough to draw up good proposals for technology education, but one that recognizes the role of the interest groups that exist in support of or in opposition to technology education. But that is only one level. As I noted at the end of the section on traditions science, CDT/ industrial arts/technical, art and design teachers will all share to some degree the characteristics of their respective traditions. Such traditions have been developing for a long time, and some have changed only slowly. As I have argued elsewhere (McCormick, 1990), teachers will need time to develop their ideas about what technology is and their skills to teach it. They will want to develop from their existing traditions, whatever mechanism for change is employed: the central dictate in the UK or the evangelism of the USA. Thus change will be a complex interplay of the national forces in professional and other interest groups, and their parallels in schools. Both levels must be attended to for the maximum effect of change.

Finally, in the context of this international workshop, we should remember that in the past there has been considerable interchange across the continents. Let us hope we can learn from history, and that we do not attempt to borrow parts of models or approaches regardless of the context within which they originated.

■ References

Cheek, D.W. (1989) Introduction. In D.W. Cheek and L.J. Waks (eds) *Technological Literacy IV*. Proceedings of the National Technological Literacy Conference. University Park, PA: National Association for Science, Technology, and Society.

Eggleston, S.J. (1976) *Developments in Design Education*. London: Open Books.

Goodson, I. (ed.) (1988) *International Perspectives in Curriculum History*. London and New York: Routledge.

Hayles, J.A. and Snyder, J.F. (1982) Jackson's Mill industrial arts curriculum theory: a base for curriculum derivation. *Man Society Technology*, 41 (5), 6–10.

International Technology Education Association [ITEA] (1988) *Technology: A National Imperative*. Reston, VA: International Technology Education Association.

Johnson, J.R. (1989) *Technology: Report of the Project 2061 Phase I Technology Panel*. Washington, DC: American Association for the Advancement of Science.

Luetkemeyer, J.F. (1985) The Social Settlement Movement and Industrial Arts Education. *Journal of Epsilon Pi Tau*, 11 (1–2), 97–103.

Marx, K. (1976) *Capital*, Vol. 1. Pelican Marx Library edition. Harmondsworth: Penguin.

McCormick, R. (1990) Technology and the National Curriculum: the creation of a 'subject' by committee? *The Curriculum Journal*, 1 (1), 39–51.

McCulloch, G., Jenkins, E. and Layton, D. (1985) *Technological Revolution? The Politics of School Science and Technology in England and Wales since 1945*. Lewes: The Falmer Press.

Medway, P. (1989) Issues in the theory and practice of technology education. *Studies in Science Education*, 16, 1–23.

Penfold, J. (1988) *Craft, Design and Technology: Past, Present and Future*. Hanley: Trentham Books.

Skilbeck, M. (1976) Ideologies and values. Unit 3, *Culture, Ideology and Knowledge*. E203, Curriculum Design and Development. Milton Keynes: The Open University Press.

1.2

Technology Education in the United Kingdom

R. McCormick

This article gives a portrayal of technological education as represented in the various proposals and curriculum documents in the UK. Scotland, England and Wales, and Northern Ireland each have their separate ways of dealing with technology education in the curriculum. This article will first outline each of the ways by exploring the place technology has in the curriculum, the aims and objectives set out for teaching it, and the structure of the resulting programmes of technology education. Following this I will consider how the documents characterize some aspects of *technology*, and how they depict the nature of technology *education*. The article is therefore not an account of current practice in schools; such an account would quickly be out of date given the state of flux in most of the UK. Even in focusing upon curriculum documents the problem of datedness exits, with current debates about 16–19 education in Scotland, possible revisions of the technology National Curriculum in England and Wales, and the fact that this article is written just before the Northern Ireland intentions become firm. The purpose of the article is less to examine what is currently happening than to consider the *kinds of decisions* that are made about technology education.

■ Outline of proposals

□ Scotland

The place of technology in the curriculum Here, the details of the technology curriculum are determined to a large extent by the public examination system, both Standard Grade (equivalent to GCSE) and Higher Grade (equivalent to A-levels). In particular, two subjects relate to technology education, *Technological Studies* and *Craft and Design*. However, there are also eight *modes of learning* and several *key skills and elements* that cut across the subject structure of the school curriculum.

The *key skills and elements* include the following process skills: technological and creative thinking (reasoning and problem solving, designing, practical applications). These skills can be developed through existing subjects and by specific courses geared to the development of the skills.

The *modes of learning* are:

- language and communication
- mathematical studies and applications
- scientific studies and applications
- social and environmental studies
- technological activities and applications
- creative and aesthetic activities
- physical education
- religious and moral education. *(SCCC, 1989)*

The aims of the *technological activities and applications* mode are to:

'. . . promote the development of technological capability through the processes of designing, making and evaluating the effectiveness of systems or artefacts; [. . .] provide opportunities for the acquisition and application of knowledge, practical skills and the generic skills of problem-solving, co-operation and enterprise; [. . .] contribute to pupils' growing awareness and appreciation of technological applications in their environment.' *(SCCC, 1989, Appendix E)*

Technological Studies and *Craft and Design* are examples of the subjects that would contribute to this mode, but there is also an identified contribution from home economics courses (see Turner *et al.*, undated).

So, although all pupils at some stages in their schooling will experience the 'technological activities and applications' mode, they will do this through various subjects which may not include either *Technological Studies* or *Craft and Design*. Consequently these subjects are not taken by all pupils at the Standard Grade level, and not all schools offer both subjects. There are also Higher Grade examinations in each of these subjects, and, although I will not describe these here, I will refer to them in the next section of the article when I consider how technology and technology education are depicted in the courses. The following paragraphs illustrate the nature of *Technological Studies*.

Aims and objectives

- to encourage the acquisition of problem-solving skills
- to develop the pupils' ability to apply the systems approach to practical problem solving

- to allow pupils to comprehend the evolutionary nature of technology and to recognize the effect of technology on the quality of life
- to highlight the role of technology in manufacturing.

Structure of the course   Technological Studies is a modular-like course. Figures 1 and 2 show that the 'areas of study' (modules) of *electronics*, *mechanisms*, *pneumatics*, and *technology and manufacture*, have permeating aspects of *systems*, *communication techniques*, *product analysis*, *energy* and *structures*. These permeating aspects are introduced in their own right (to provide a basis for use) and are also developed within the modules as appropriate. (Figure 1 also shows that the modules are to be interrelated as appropriate.)

The Introductory Unit (Figure 2) should have a duration of about 40 hours, and is designed as a 'taster unit' which will, in addition to introducing pupils to the main areas of study, provide an introduction to the systems approach and problem solving. Initially, pupils should be introduced to problem solving in the modules (electronics etc.), but as the course progresses, the realistic integration of these areas will become apparent. The course is based upon problem-solving activities, but there is a clear delineation of knowledge and understanding that is required to be achieved by pupils.

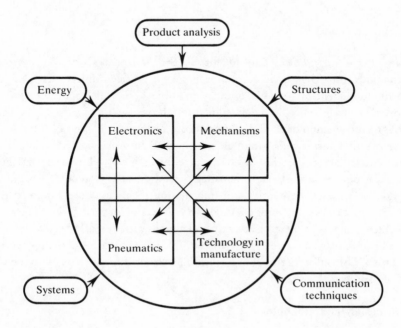

Figure 1

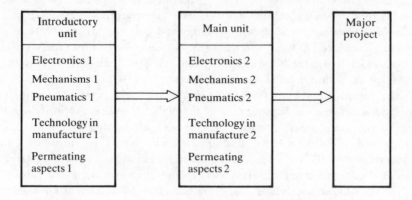

Introductory unit	Main unit	Major project
Electronics 1	Electronics 2	
Mechanisms 1	Mechanisms 2	
Pneumatics 1	Pneumatics 2	
Technology in manufacture 1	Technology in manufacture 2	
Permeating aspects 1	Permeating aspects 2	

Figure 2

☐ **England and Wales**

Place in the curriculum   The advent of the National Curriculum brought technology into the curriculum, but not as a unified subject. The National Curriculum identified it as a 'foundation subject', not as important as a 'core subject' such as science, and defined it through two distinct profile components, 'design and technology' and 'information technology'. For all intents and purposes these are two separate elements, with information technology being cross-curricular, to be developed in all subject areas. However, the former subjects of art and design, business studies, CDT (craft, design and technology), home economics and information technology are supposed to be co-ordinated to present integrated design and technology activities. The degree to which teachers of these subjects do collaborate varies greatly between schools. At the time of writing (early 1992) the link between these existing 'subjects' and the GCSE subjects that will be created was not direct. There are to be three GCSE subjects:

- design and technology
- information systems
- technology.

Aims and objectives The whole of the National Curriculum is defined in terms of Attainment Targets (ATs) with the various subjects having varying numbers of them (e.g. in mid-1992 science has four), and these are then broken down into more detailed Statements of Attainment (SoA). The ATs define the aims of the subject in terms of what pupils will be able to do. Technology is made up of two sets of ATs: ATs 1–4 for design and technology capability, and AT 5 for information technology capability. The ATs for design and technology are:

- AT 1 *Identifying needs and opportunities* Identify and state clearly needs and opportunities for design and technological activities through investigation of the contexts of home, school, recreation, community, business and industry.
- AT 2 *Generating a design* Generate a design specification, explore ideas to produce a design proposal and develop it into a realistic, appropriate and achievable design.
- AT 3 *Planning and making* Make artefacts, systems and environments, preparing and working to a plan and identifying, managing and using appropriate resources, including knowledge and processes.
- AT 4 *Evaluating* Develop, communicate and act upon evaluation of the processes, products and effects of their design and technological activities and of those of others, including those from other times and cultures.

AT 5, information technology (IT) capability, requires pupils to be able to use IT to: communicate and handle information; design, develop, explore and evaluate models of real or imaginary situations; measure and control physical variables and movement.

Structure of the programmes The Programmes of Study (PoS) are intended to describe the content, skills and processes which are to be covered during each of four key stages (defined to be reached by pupils at ages 7, 11, 14 and 16). The PoS assume 'design and make' activities are basic to the teaching and are built around four headings:

- developing and using artefacts, systems and environments
- working with materials
- developing and communicating ideas
- satisfying needs and addressing opportunities.

Very little content is specified directly, although it is possible to see some of the remnants of subject areas of an earlier report (DES/WO, 1989), for example: energy, structures, mechanisms, systems. Now, however, they are scattered among the four headings and the statements in the PoS resemble the SoA in their form and sometimes in their content. For example, a PoS statement:

allocate time and other resources effectively throughout the activity

and a SoA:

consider constraints of time and availability of resources in planning and making.

☐ Northern Ireland

Place in the curriculum   As in England and Wales, the Northern Ireland Curriculum has a subject which provides technology education: *Technology and Design*, giving another variation on the title used in England and Wales! Although links are envisaged between *Technology and Design* and other subjects such as art and design, home economics, and information technology (this is a cross-curricular theme), they have their separate existences (Ministerial Technology and Design Working Group, 1991). *Technology and Design* is recognized as being found in the traditional subjects of CDT, home economics, and art and design.

Aims and objectives   The general aim of *Technology and Design* is:

'. . . to enable pupils to become confident and responsible in solving real life problems, striving for creative solutions, independent learning, product excellence and social consciousness.' (*Ministerial Technology and Design Working Group, 1991, p. 15*)

This translates into the more specific aims to:

- stimulate pupils' curiosity, imagination, creativity, and to develop the ability to operate effectively in society
- to develop, through active learning, pupils' understanding of technology and design
- develop pupils' abilities to identify and respond to needs, opportunities, 'technology push' and 'market pull'
- involve pupils in purposeful design activities resulting in the development of products
- promote the ability in pupils to communicate information and ideas through a variety of media
- equip pupils with technology and design skills to enable them to cope with and respond, more effectively to an increasingly technological society
- develop pupils' awareness of the nature and application of technological products
- develop an appreciation of the importance of quality

- promote an awareness of the means by which societies generate wealth
- enable pupils to appreciate the importance of technology and design in society, historically and in the present day, particularly as it affects the economy
- raise pupils' awareness of the implications of technology and design decisions on society and the environment
- engage in the interests of all pupils and help sustain their motivation for learning
- promote awareness of health and safety issues relating to technology and design.

At the proposal stage there was one attainment target – technology and design capability. To develop this capability, pupils should acquire, develop and apply in parallel:

- scientific knowledge and understanding
- a range of intellectual skills
- a range of physical skills
- a range of communication skills
- an awareness of the implications of technology and design on the community, economy and environment.

The above should be developed principally through the design and manufacture of products.

Structure of the course As is the case in England and Wales there are Programmes of Study (PoS) and these at times resemble objectives (e.g. select, and use safely, tools appropriate to the task). In Key Stages 3 and 4, three broad headings are used: knowledge, designing and communicating, and manufacturing. These PoS only define what is done in broad terms, and the 'design and manufacture' activities indicated by the AT will dominate the proposed courses. These activities involve, at different levels to match the Key Stage of the pupil:

- investigations of contexts and products
- discussing tasks, needs or opportunities
- modelling in two dimensions to aid design thinking
- communicating including through IT
- making/manufacturing of products in a range of materials including wood, metal, plastics, and textiles

- using electronic, mechanical and pneumatic components in these products.

Specific reference is made to the Northern Ireland science ATs including the content areas of energy, materials, forces, electricity and magnetism, and sound, light and waves. Some of these areas are also mentioned in the statements of knowledge in *Technology and Design*, along with: control and systems ideas; the operation of devices (e.g. switches and sensors); use of tools and machines; health and safety at work; standards and regulations covering products. (These vary in the way they are worded for the various Key Stages.)

■ Views of technology

Here I want to look at how a number of important features of technology are treated in the various curricula. In particular I will consider how wide is the range of activities that are included, the treatment of values, the definition of technological content and the technological processes that are included. (See McCormick (1990) for a fuller list of such features.)

☐ The range of activities

The Scottish *Technological Studies* is quite narrow in its coverage of areas of technology, as indicated by the modules (electronics, mechanisms, etc.), whereas England and Wales and Northern Ireland have in theory much less restriction. However, the Higher Grade *Technological Studies* includes an industrial study that requires students to investigate a whole range of aspects. These aspects include the usual technical considerations about the processes, products and instrumentation, but also social implications, marketing, economics and environmental impact. The contexts defined in England and Wales and Northern Ireland (home, leisure, industry, etc.) offer a wide scope for examining technology in a variety of settings. The limitations are more likely to come from the knowledge that the teachers possess and the difficulties of representing some kinds of technology in schools. This latter limitation is exacerbated by the preoccupation of some teachers and syllabuses with 'design and make' activities, and the consequent underplaying of investigations, simulations and discussions that have no practical (product) outcome.

☐ Value judgements

Again Northern Ireland and England and Wales proposals are similar in the way they deal with value judgements. They are usually dealt with in relation to pupils' 'design and make' activities, through considering for example the social, environmental and economic implications of products (this is particularly so in the SoA). Northern Ireland *Technology and Design* has statements in the proposed PoS that indicate a concern for value judgements, for example:

> '. . . consider that the use and manufacture of products can have social, economic and environmental implications . . .' (*Ministerial Technology and Design Working Group, 1991, p. 28*)

In contrast *Design and Technology* in England and Wales has this kind of statement in the SoA (AT 4, SoA 4d):

> '. . . understand the social and economic implications of some artefacts, systems or environments . . .' (*DES/WO, 1990, p. 16*)

What is less common is asking fundamental questions about the nature of technology or about the desirability of a particular technological activity. However, there is an awareness of the need to be questioning (for example in AT 1, SoA 10a):

> '. . . convey [. . .] that their [pupils'] identification of needs and opportunities is justified and worth developing . . .' (*DES/WO, 1990, p. 6*)

Both courses also recognize the culture-specific nature of some technology, for example in England and Wales (AT 4, SoA 6e):

> '. . . illustrate the economic, moral, social and environmental consequences of design and technology innovations including some from the past and other cultures . . .' (*DES/WO, 1990, p. 17*)

Unfortunately such statements are scattered throughout the SoA and have to be dug out to see the role of value judgements, which may not therefore be consistently treated by teachers. The Northern Ireland proposals contain aims directed towards value judgements (as was some of the discussion in the England and Wales proposals – DES/WO, 1989, p. 9), but when the Northern Ireland proposals get into the form of legislation these aims may similarly be buried in the details of a few SoA.

In Scotland there are few indications of value judgements being important, except in a rather biased way in Higher Grade *Technological Studies*. In describing a framework for pupil-centred learning, pupils are seen to start by asking the question 'How does technology help mankind [*sic*]?' (Scottish Examination Board (SEB), 1990, p. 28). Investigating this question through the various elements of the course is seen only in a positive light, i.e. technology is always positive. While this may not have been the intention of the course designers, not making explicit the importance of value judgements runs the risk of teachers ignoring them, underplaying them or only seeing technology as positive.

☐   The definition of content

Scotland, through the 'areas of study' (the modules on electronics, mechanisms, etc.), has a quite clear statement of content, even to the extent of listing the terminology and concepts that pupils have to understand. There is therefore no doubt about what the problem solving will be based upon. In contrast in England and Wales the content is not specified clearly. The SoA are 'process-based' (dealing with processes such as planning, investigating). Because of the importance of the SoA in defining what is assessed (and hence in the teachers' minds what *must* be done), the process tends to dominate, with teachers being unsure of what content should be covered. Some see the content only being defined by the medium of the work (e.g. wood, metal, food, textiles), but are not sure how much of energy, control, structures, mechanisms, or electronics, for example, should be dealt with. This uncertainty is part of the motive for revision of the curriculum I indicated in the introduction to the article.

The Northern Ireland Curriculum for *Technology and Design* is again much the same as that in England and Wales, except that there is a listing of the knowledge required to be understood by pupils. There is even a categorization into different kinds of knowledge (applicable, general and contextual), but within the PoS there is simply a list of items with no evident structure, and no identified progression from one Key Stage to another.

☐   The selection of processes

The three 'national curricula' include a variety of technological processes. Among these processes are designing, problem solving, modelling, making/manufacturing and the systems approach. I will comment on only some of these.

The Northern Ireland proposals and the National Curriculum of England and Wales both emphasize designing and making as the core processes, in contrast to Scotland where *Technological Studies* is based upon a problem-solving process. For England and Wales the ATs represent the abilities to be fostered in designing and making, and in Scotland there are similar 'sub-elements' of problem solving: analysis of the problem, performance criteria, ideas for solutions, selection of resources, building and testing, modification, evaluation. However, there is a rather confusing link made with the systems approach, where problem solving is defined as 'the practical application of a systems approach' (SEB, 1989, p. 8). Elsewhere the systems approach is discussed in terms of representing something in terms of systems and sub-systems (the latter being represented by an input, process and output), in other words as a *method of analysis.* Given their definition of problem-solving, the systems approach would only be part of problem solving.

Because of the separation of *Technological Studies* and *Craft and Design*, there is a danger of designing being separated from *Technological Studies*, but this is not the intention. The use of the term 'problem solving' does in theory allow processes other than designing to be considered. In both Northern Ireland and England and Wales design dominates, not surprisingly given the titles of the curriculum areas. But this means that processes such as *repairing* and *using* technology are ignored, and in England and Wales *manufacturing* is also underplayed.

The fact that processes such as repairing are ignored can lead to some bizarre tasks being set. In *Technological Studies*, for example, there is a specimen examination question where pupils are to design a detector for indicating when a washing machine is leaking, when the obvious solution is not a moisture detector, but to *repair* the washing machine. What is to be done when the detector indicates a leak: turn off the machine, stand and watch, or simply mop up (and risk electrocution)? Of course the point of the question is to test the students' understanding of moisture-detecting circuits, but the hidden curriculum is also powerful! (The situation is made worse by the fact that the question completely ignores how such a detector might be used, with no account being taken of the leak varying in amount and hence position, or whether the lead from the sensor would have to trail across the floor, or the whole circuit would get flooded and be useless.)

The dominance of design, in my view, misrepresents the nature of technological activity as it occurs in the world outside school, where only a few per cent of those involved in technology would be involved in design. Of course some would argue that the title of the curriculum area (Design and Technology) justifies the emphasis. But then the danger exists of creating a 'subject' that has no identity outside school, and gives pupils a false view of the culture of which they are part. (Alternatively they will recognize the mismatch with technology as they experience it outside school and give it a low status in their thinking, as it has had in the thinking of the nation. This has been an enduring problem for practical activity for a long time (see McCulloch, Jenkins and Layton, 1985).

■ Views on technological education

Not surprisingly, the views of technology discussed above influence those of technology education. The development of technological capability and awareness are the central features of technology education in the UK. There is some argument about whether or not they should be separately identified. In an earlier discussion document in Scotland their separation was reinforced by each being seen to be developed through different courses: capability developed through *doing* technological activity; awareness through courses *about* technology, e.g. in social studies (CCC, 1985, pp. 10–11). In fact *capability* is the dominant feature in the curricula I have been discussing. *Awareness* is included within capability, covered through evaluation activities, and in identifying needs and opportunities for technological activities (as I indicated when considering value judgements earlier).

The views on technological processes discussed above also reflect the way capability is defined. Thus design-and-make is the dominant feature of capability. In England and Wales this is reflected in the ATs, which ignore abilities to diagnose and repair faults or even to use technology. Using technology is arguably the most common encounter anyone is likely to have with technology. As I have argued elsewhere (McCormick, 1990, p. 43), some of the other elements of capability were identified in the original proposals for England and Wales, which were ignored in the final National Curriculum document. These proposals listed the following as being part of capability:

- pupils are able to *use* existing artefacts and systems effectively
- pupils are able to make *critical appraisals* of personal, social, economic and environmental implications of artefacts and systems
- pupils are able to *improve*, and extend the uses of, existing artefacts and systems
- pupils are able to *design*, *make* and *appraise* new artefacts and systems
- pupils are able to *diagnose* and *rectify faults* in artefacts and systems. (*DES/WO, 1988, pp. 17–18; my emphasis*.)

The Scottish *Technological Studies*, with its use of problem solving in theory, overcomes this dominance of 'design-and-make', although not necessarily in practice. In fact this course defines three assessable elements of the course, which could be interpreted as its view of capability:

- knowledge and understanding – relevant and applied to problem-solving situations
- technological communication – verbal and graphical
- problem solving – logical sequences of systems approach.

■ Conclusion

It is surprising that in such a small country there can be so much variation in technology education. There are of course some common features, more between Northern Ireland and England and Wales, than between them and Scotland. In part this is because Scotland was quicker off the mark and moved into *Technological Studies* at a time when modular courses similar to this course were common in CDT versions of technology in England and Wales. It is interesting to see how their views of technology are different, when they are all trying to represent to children the same part of human culture! The important lesson to draw from this discussion is to see all the curricular statements as representing views at a particular period, views that will have to be developed. Also they indicate the importance of teachers both understanding the rationale underlying them, and being prepared to develop them.

■ References

Consultative Committee on the Curriculum (CCC) (1985) *The Place of Technology in the Secondary Curriculum. Final Report of the CCC's Committee on Technology*. Dundee: Dundee College of Education or the CCC.

Department of Education and Science and the Welsh Office (DES/WO) (1988) *National Curriculum Design and Technology Working Group Interim Report*. London: DES/WO.

Department of Education and Science and the Welsh Office (DES/WO) (1989) *Design and Technology for Ages 5 to 16: Proposals of the Secretary of State for Education and Science and the Secretary of State for Wales*. London: DES/WO.

Department of Education and Science and the Welsh Office (DES/WO) (1990) *Technology in the National Curriculum*. London: HMSO.

McCulloch, G., Jenkins, E. and Layton, D. (1985) *Technological Revolution? The Politics of School Science and Technology in England and Wales since 1945*. London: Falmer Press.

McCormick, R. (1990) Technology and the National Curriculum: the creation of a 'subject' by committee? *The Curriculum Journal*, 1 (1), 39–51.

Ministerial Technology and Design Working Group (1991) *Proposals for Technology and Design in the Northern Ireland Curriculum*. Report of the Ministerial Technology and Design Working Group. Belfast: Northern Ireland Curriculum Council.

Scottish Consultative Council on the Curriculum (SCCC) (1989) *Curriculum Design for the Secondary Stages*, (Guidelines for Headteachers), 1st revised edition. Edinburgh: SCCC.

Scottish Examination Board (SEB) (1989) *Standard Grade Arrangements in Technological Studies*. Edinburgh: SEB.

Scottish Examination Board (SEB) (1990) *Arrangements for Higher Grade Technological Studies*. Edinburgh: SEB.

Turner, E., Black, H., Hall, J. and Devine, M. (undated) *Technology in Home Economics*. Edinburgh: Scottish Council for Research in Education.

1.3

Technology Education in the Netherlands

M.J. de Vries

■ **A short history of technology education in Dutch schools**

To understand the Dutch educational situation, it is necessary to realize that secondary education is split up in two parts: general education and vocational education. That means that pupils after eight years of primary education (from ages 4 to 12 years) have to choose either to go to secondary general education or to vocational education. In general education they can choose between a six year course (pre-university), a five year course (pre-higher vocational), and a four year course (pre-middle vocational). In the vocational area there are several types of (lower) vocational courses, such as technical, agricultural, economical and administrative. These are all four year courses. Figure 1 gives a scheme of the complete Dutch education system.

The vocational courses are generally seen as for less able pupils, although this was never the intention of these types of schools. The choice of the pupils in most cases is guided by the results of a test at the end of primary education and the headteachers' advice. For many pupils lower vocational education is their final education before they get a job. Others may go on to middle vocational education.

In 1973 in the vocational schools a subject called 'General Techniques' was introduced. Originally it was aimed at giving pupils a general introduction to the world of technology. As no curriculum was provided, and 'technology' was not defined, the subject was given all kinds of meanings, except for the one that was intended originally. In most cases, all that was taught was the practical handicraft skills that pupils needed for the rest of their vocational education. Those skills differed greatly between the various types of vocational schools. In technical schools, for example, it would be woodwork and metalwork, in schools of economics, it would be an introductory book-keeping course.

The subject was taught by teachers who were educated to teach a subject

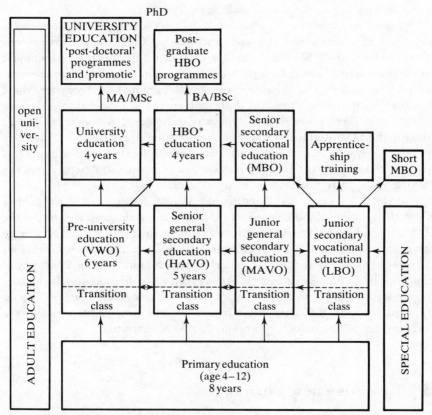

Figure 1   The Dutch education system.

*Note: Dutch government publications in English refer to HBO (hoger beroepsonderwijs) as Higher Vocational Education. The HBO institutes themselves prefer the term Higher Professional Education

that was seen as related to technology (e.g. woodwork, physics). To help the teacher 'survive' in the subject 'General Techniques', inservice courses were provided by teacher-training institutes. These courses were relatively short and it was not compulsory for a teacher to take such a course. In many cases the teacher taught a subject that resembled his original subject as closely as possible. This could be seen from the names that were given to the subject: 'General Techniques/Wood', or 'General Techniques/Metal'.

The history of technology education in Dutch schools is to a large extent dominated by this craft origin. UK readers will recognize this, as the origin of 'Design and Technology' also stems from a craft-like subject, out of which grew 'Craft, Design and Technology'. An important difference, however, is that in the Netherlands the subject in its initial state was completely developed in the vocational area. The design aspect therefore came in much later and even today is not accepted generally, as I shall show later on.

To get more coherence in various 'General Techniques' between schools and types of schools, the National Institute for Curriculum Development (Dutch abbreviation: SLO) published a document, in which 63 attainment targets were stated, based on a philosophy that emphasized the human–technology relationship. The attainment targets were grouped in three parts: those that dealt with matter, those that dealt with energy and those that dealt with information. These were called the *pillars* of technology. A number of schools adopted the SLO attainment targets, but the subject did not have the same content nationwide and the craft skills kept dominating the subject. In part this was due to the fact that the curriculum document was not so different from the existing craft-skills teaching in many schools.

In 1985 a decision was taken to take all *care* elements out of the subject 'General Techniques' and to put them into a separate subject. This forced the SLO to publish a new document that took this decision into account. But from this document it can be seen that other changes had also taken place. Gradually, the movement was away from the 'pillars' approach to looking for a new way of grouping the attainment targets. What emerged was a division into three parts: technology and society; dealing with technical products; and developing technical products. Later on, this division would be quite influential in the development of technology education in the Netherlands.

■ Towards a new situation

In 1985, not long after the decision to cut the 'care' elements out of 'General Techniques', the government made a proposal to parliament to change the content of the first three years of secondary education. The motive for this was that many people were unhappy with the fact that pupils at the age of 12 were forced to make a decision between general education and vocational education. This decision was particularly important, because it almost never arose that a pupil would switch from general education to vocational education, or vice versa. Therefore this decision, taken at such an early age, would to a large extent determine the future career of a pupil. The idea of the government proposal was to make the curriculum in general and vocational education more parallel, so that pupils could move more easily from one type to another. It was proposed to make a curriculum of 14 subjects (later, this was increased to 15 subjects). For each subject a committee was formed to develop attainment targets for that subject. In many respects it resembled the National Curriculum in the UK, except for the fact that in the Netherlands there was already a tradition of nationally established exams instead of regional exam syllabuses. All schools, both in the general and in the vocational area, now have to prepare for the introduction of the new curriculum in 1993.

We will now take a closer look at the content of technology education as one of the 15 subjects of the new curriculum. As the only official document available is the list of attainment targets, we will take that list as a starting point. The list was developed by one of the attainment target groups which consisted of representatives from secondary schools, teacher-training institutes, the National Institute for Curriculum Development, industry, the Central Institute for Test Development (Dutch abbreviation: CITO) and the commercial publishers. The total amount of teaching hours for technology education is 180 (according to a table that has an advisory status: schools can change that number as long as they deal with all attainment targets in the subject). In practice that means that technology education can be taught two hours a week for 2½ years.

The introduction to the list of attainment targets states that:

'technology education is aimed at getting pupils acquainted with those aspects of technology that are important for a good understanding of culture, for functioning in society and for further technological schooling. Pupils acquire knowledge and insight into the three pillars of technology (matter, energy and information), the close relationship between science and technology and between society and technology. They learn to develop technology, acting in a practical way, to deal with a number of technical products and they get the opportunity to find out their own possibilities and interests with respect to technology. In its elaboration, technology education should be attractive and meaningful to both girls and boys.' (Ministeric van Onderwijsen Wetenschappen (1990) *Advics kerndoelev voor de Basisvorming*, Der Haag, Stastsuit-gevarg.)

The attainment targets were divided into three *domains*:

- A   Technology and society
- B   Dealing with technical products
- C   Making functional workpieces

For each of these domains a number of 'subdomains' was formulated, each of which was related to one attainment target. For domain A the subdomains (and related attainment targets) were:

(1) *Daily life* – pupils can state the consequences of post-war technology for society in the areas of living environment, work and careers, leisure activities, and traffic.

(2) *Industry* – pupils can describe the functioning of a production industry after having observed it.

(3) *Professions* – based on concrete information, pupils can indicate to what

extent professions make use of technical products and how these professions are changed by that.

(4) *Environment* – pupils can give examples of the influence of technological developments on the environment and show how pollution can be avoided or diminished by technological applications.

For domain B the following subdomains were formulated:

(5) *Working principles* – pupils can indicate the nature of transmissions and motions in a concrete situation and choose a suitable transmission for a given situation and construct this with given elements.

(6) *Technical systems* – pupils can indicate the energy conversions in the petrol engine and electrical engine, dynamo, central heating installation and the sun panel; they can indicate the function and application of materials and use static construction principles and connection means in models.

(7) *Control technology* – pupils can indicate the relevant elements of a control system and recognize mechanical, electrical, and pneumatical control; they can make a working model of a control system and globally indicate the principle of a level or temperature control system; they can connect and use computer-controlled systems.

(8) *Using technical products* – pupils can use the correct terms when using technical products, use the correct tools, apparatuses and means and maintain them; they can act according to given instructions for assembly and use, and evaluate products in their own environment with respect to safety and functionality; they can indicate the functions of parts in technical products as far as they are relevant for maintenance and repairing.

The subdomains of domain C were:

(9) *Preparation of work* – pupils can use technical documentation and instructions, develop a working plan and interpret safety instructions.

(10) *Design, making and reading technical drawings* – pupils can make a sketch of a workpiece, to be made by themselves, read a technical drawing with symbols in it, and do necessary measurements and transfer them to materials.

(11) *Working with materials* – pupils can separate, reshape, connect and combine the materials wood, plastics, textiles and metals, and make a workpiece according to an incomplete instruction and according to a model for problem solving; they can cooperate to make a product in a series. To that aim they can set up a production line and, within a certain time limit, carry out the partial actions, assemble the product and control them.

(12) *Control of workpieces* – pupils can evaluate workpieces and the process to

make them according to criteria for the use of materials, usability, completing, environment, safety.

The meaning of these attainment targets for technology education practice can be seen from the example materials that the SLO has produced to illustrate them.

The SLO produced a curriculum document in which they propose two different plans for themes that together fill the 180 hours for technology education. These are just examples; schools have the right to decide whether or not to use them. The only things schools are forced to do is to make sure that pupils learn the attainment targets. The first plan of the SLO contains 12 themes, divided over two years. In the first year they have the following themes:

(1) Introduction to the subject, the classroom and the working procedures.

(2) Maintaining your own bike.

(3) Technology and housing.

(4) A complex workpiece.

(5) Craft and machines.

(6) Various issues and/or time for pupils who did not complete all earlier themes.

The second year has the following themes:

(1) Introduction to construction systems (e.g. Lego, Fischer).

(2) Levers and pulleys.

(3) Process technology with Lego robotics.

(4) Technology and drinking water supply.

(5) Energy in technology.

(6) Complex workpiece.

The second plan contains six themes for two years:

(1) Introduction to the subject and the classroom, safety, and basic skills.

(2) Introduction to computer use, mechanisms and systems.

(3) Practical assignments on electricity in our homes.

(4) Technology in my own room.

(5) Production lines, mass production.

(6)   Practical assignments on water in the home, heating, the bike, cleaning, clothing.

When comparing these *themes* to the current situation of 'General Techniques', it is striking that the differences are small. There is a bit more advanced technology in the curriculum, there is a (small) amount of hours dedicated to the relationship between technology and society, and the process of making workpieces is somewhat more open than before, although many decisions have still been taken before the pupils enter the process. In that sense, certainly, design is still not an important issue in technology education. This in spite of the fact that the SLO formulated a procedure, called thinking–drawing–making–control (Dutch abbreviation: DTMC), in which design is mentioned.

Perhaps the major difference is that 'General Techniques' was completely structured according to groups of skills (a kind of 'systematic' approach: a structure, that was derived from the discipline) and now the structure is a mixture of 'systematic' and 'contextual'.

■ Developing a framework

It will probably be evident to UK readers that there is not much coherence in the Dutch attainment targets. Not only is there a lack of continuity from one attainment target to the next but also, within attainment targets, it often seems that issues have been selected randomly from many other possibilities. Some examples:

- in attainment target 1, the history of technology is limited to post-war history only. The choice of contexts seems to be random (e.g. why is communication absent, why limit transport to traffic?);
- in attainment target 6, pupils seemingly do not have to know what a system is and what its essential elements are; there is no awareness that this is where matter, energy and information come together. Instead, a random choice of technical devices is given, of which pupils only have to recognize energy conversions and the use of materials;
- the selection of materials in attainment target 11 is small and hard to defend when one realizes that many of the materials pupils see around them do not belong to these (e.g. glass, bricks, paper, ceramics);
- in domain C, design as a process of stating requirements, developing alternatives, evaluating and choosing, seems to be absent completely. This cuts out the very heart of the technological process.

The shift from systematic to contextual had, as a consequence, the emphasis placed on certain specific fields in which technological products are used. These products change quite rapidly and a technology subject that only focuses on these products only will need very frequent updating to prevent it from becoming outdated.

Many more comments can be made on the list. One of the most important criticisms is that the whole idea of product *development and innovation* – a key issue in technology as it is practised in an industrial context – is absent. Here it is striking that industry in the Netherlands clearly has not 'discovered' the subject technology yet. Although industry was represented in the attainment target group, no serious attempt was made to include product innovation in the attainment targets. Here again can be seen a striking difference with the UK situation, where school–industry links are much more common for technology education than in the Netherlands. In the Netherlands, industry seems to be only interested in contacts with vocational schools and general education is still outside their scope of interest.

From these points of criticism (no coherence, absence of product innovation) grew the need for some people involved in the development of teacher education programmes in the Netherlands to create a conceptual framework. This framework would put all the attainment targets into a context that would help teachers and pupils recognize the relationships between these targets and thus help them to acquire a well-balanced concept of technology as a discipline and as a cultural phenomenon. In particular, this was done in the Pedagogical Technological College (Dutch abbreviation: PTH), located in Eindhoven (a city well known because of the activities of Philips). This institute had served as a hosting institute for a series of international conferences on technology education (the so-called PATT conferences: Pupils' Attitude Towards Technology). These conferences brought together technology education specialists from many countries worldwide. These contacts were a great help in establishing a philosophy of technology education and technology education teacher-training for the PTH. Elements from various conceptual models for technology could be combined into a new model, as can be seen in Figure 2. We will now explore this model and its meaning.

The main characteristic of the model is the vertical line of designing–making–using. This line indicates the process through which technological innovations take place. At the left side, the inputs for each step are made visible, at the right side, the outcomes of each step can be seen.

The process of designing is initiated by three elements that have been drawn in the top left part of the model: human needs, norms and values, and scientific knowledge. The human needs lead to technological problems. Norms and values to a certain extent determine if and how we will use technology to accomplish our needs. Scientific knowledge serves as an input of ideas and sets limitations to the solutions we can look for in the design process.

As can be seen in the model, the making process is seen as a conversion of material, energy and information into other forms of material, energy and information, that we call 'product' and 'waste'. In essence the same happens when we open the box 'using', but we did not choose this one because one of the

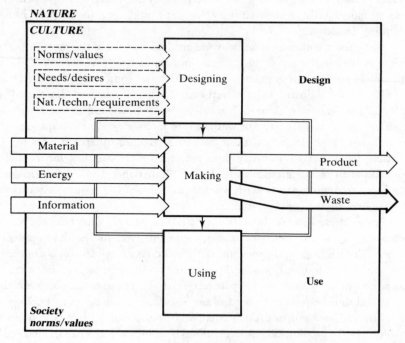

Figure 2 The PTH framework for technology.

three (matter, energy and information) may be absent here (a TV set does not convert matter, only energy and information.

The using phase does not only refer to dealing with products as consumers, but also maintenance and repair of products, and ultimately disposal or recycling, i.e. essentially the categories we may find in a manual. The using process brings us into society, where further norms and values to a certain extent determine how technological products are used.

The teacher-training programme in Eindhoven has been elaborated from this model. Some parts of the programme deal with the model as a whole (e.g. Philosophy of Technology, Technological Principles), other parts of the programme deal with only parts of it (e.g. Sketching/Drawing/CAD, Materials, Workshop Practice). Important parts of the programme are the projects, in which student teachers practise the model by choosing a technological problem, designing a solution, making it and thinking about its use and possible role in society. In these projects, student teachers are forced to consider the use of technological concepts and principles (the concept of 'systems', morphology and *synectics*, the use of function–form–materials–treatment relationships, illustrated by the so-called 'function triangle' – see Figure 3). These are a few examples of a whole set of possible concepts and principles. A lot of development work still needs to be done to create a complete and systematic set of technological concepts and principles. The student teachers must also think about the possible use of knowledge from science.

■ Two streams

As the curriculum proposal does not change the educational structure of secondary education in the Netherlands, there is a danger that two streams of technology education will emerge: a vocational stream that will continue the 'General Techniques' tradition, and a general stream that will try to implement a completely new subject, Technology Education. It can already be seen that commercial publishers develop separate materials for general and vocational schools. Characteristics for the vocational stream will be:

- practical skills as the main aim;
- emphasis on making and using skills;
- relatively small attention to conceptualization;
- hardly any relationship with science and maths;

whereas for the general stream they will be:

- practical skills as both an aim in itself and a means for teaching technological concepts;
- a balance between designing, making and using;
- working towards formation of a balanced concept of technology with the pupils;

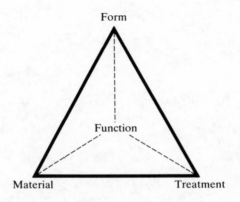

Figure 3 The 'function triangle'.

- teaching basic technological concepts and principles through practical activities;
- an attempt to show how knowledge and methods from mathematics and science are used in technology.

This danger has been recognized by the SLO and they are now revising the curriculum document to make it fit the needs of both vocational and general schools. More attention will be paid to conceptualization. Hopefully the SLO will be able to prevent a two-stream situation emerging and challenge people in vocational education to rethink the content of their technology education courses and take into account recent developments, as discussed at an international level. Another influential factor in further developments will be the commercial publishers. Textbooks can have a strong impact on the content and method of school subjects. Some publishers have explicitly stated that they are working on a method that will suit the purposes of the general schools. Within a few years, probably four or five of them will bring a method for technology education in general education on the market. At this moment it is hard to predict which stream will be chosen by the majority of schools. Most schools have just started orienting themselves on the content and boundary conditions of the new subject technology. There is a lot of confusion among teachers, most feeling hardly competent to teach this new subject, either because they become eligible to teach it just because of their 'General Techniques' experience, or because they went through the extremely short and condensed retraining programme. It can be expected that the coming years will be crucial for the further development of technology education in the Netherlands.

1.4

Technology Education Proposals in the USA

R. McCormick

■ Introduction

This article starts by briefly considering the history of developments that led to industrial arts teaching (equivalent to the British craft tradition) and thence to technology education. It then moves on to consider proposals for technology education from three recent projects, two from the science fraternity (*Educating Americans for the 21st Century*, and *Project 2061*) and one from technology educators (*A Conceptual Framework for Technology Education*). As in the previous article on technology education in the UK (Article 1.2), I will concentrate on curriculum proposals. The first set of proposals (*Educating Americans for the 21st Century*) was developed in 1983 and has not resulted in any impact upon technology educators. *Project 2061* has fostered a considerable development programme, but the extent of the involvement of the mainstream of technology education teachers is unclear. The *Conceptual Framework* has only recently been developed and as yet has not influenced practice. It is important to realize that all the proposals are largely the result of the academic community's discussions. As Booth (1990) reports, in the USA there is a wider gap between such proposals and the reality of schools than would be the case in the UK, where teacher involvement in curriculum developments is more common.

■ Historical background

The USA was influenced by both Russian manual training and Swedish *sloyd* during the late 19th century and early 20th century, just as the UK was (Hacker and Barden, 1983; Eyestone 1989). At the beginning of the 20th century, Richards (at the Teachers College, University of Columbia) built on Dewey's work by using a socioeconomic analysis to identify *manufacturing*, *transportation* and *communication* as areas of what became the industrial arts curriculum (Luetkemeyer, 1985). This provided the basis, almost 80 years later, for the Jackson's Mill Curriculum Project's analysis of industrial arts as the study of the human adaptive systems of *manufacturing*, *construction*, *communication* and *transportation* (Snyder and Hales, 1981). Industrial arts was eventually renamed as technology education and has, in some forms, developed a problem-solving approach. The 'design and make' developments in the UK have had some influence upon the USA, but there are still many examples of traditional craft approaches, often in a vocational context. Even the study of the 'human adaptive systems' can be very theoretical, requiring very little by way of practical action. In this sense, the USA technology educators have always been better at defining the content of technology than its processes.

■ Educating Americans for the 21st century

This was a project set up by the National Science Foundation in an effort to regain the USA's technology lead over the rest of the world. As an initiative of a science body, it not surprisingly suggested a strong integration of science and technology, and indeed mathematics. The Commission that made the proposals was concerned with the relatively poor performance of the bulk of pupils in the USA compared with Japan. As is the case in the UK the best in the USA compare well with those in Japan, but the 90% below the best do much worse than those in Japan. This kind of analysis led the Commission to propose not a national curriculum, but a framework within which alternative curricula and materials could be developed.

The features of their proposals were:

- more integration of science, mathematics and technology
- a concern for education for citizenship as well as for those who go on to be engineers and scientists
- an increase in the amount of science, mathematics and technology taught
- a concern to increase the motivation of students to study science, mathematics and technology. (*NSF, 1983a, p. 44*)

The aims for science and technology (mathematics is defined separately) were:

- to deal with the complex social issues to be addressed by citizens
- to enable students to understand technological innovation, the productivity of technology, and the impact of products and the social consequences.

The objectives for students were:

(1)   to formulate questions about nature

(2)   to have a capacity for problem solving and critical thinking

(3)   to have a talent for innovative and creative thinking

(4)   to be aware of science and technology careers

(5)   to have the basic academic knowledge for advanced study

(6)   to have scientific and technical knowledge for civic responsibility and for their own life

(7)   to be able to judge the worth of articles presenting scientific conclusions. (*NSF, 1983a, p. 44*)

They also wanted students to be able to use the knowledge and products of science, mathematics and technology in their lives and work, and to make informed choices in their personal lives and in social and political arenas.

On the face of it, the proposals may seem to be typical of the science and STS traditions (see Article 1.1), with little understanding of the uniqueness of technology. However, the proposals and the source materials upon which they were based (NSF, 1983b) showed a sophisticated (but not always consistent) view of both the nature of technology and the learning of technological processes. For example, there was a strong emphasis upon practical issues and dealing with practical problems. This was partly as a motivational device to encourage the learning of the scientific concepts, but also to develop problem-solving skills. The recognition of the uniqueness of technology was evident in the delineation of technological concepts and skills, and in the recognition of the importance of synthesis and design. The design process was recognized as ultimately producing technological change and as being governed by economic and social considerations (NSF, 1983b, pp. 61–2). It was also recognized that synthesis and design were missing from science and mathematics courses and needed additional topics in science and technology courses.

The definition of technology used within the proposals contains some interesting phrases indicating the close link of science, mathematics and technology:

> '. . . the tools, devices and techniques that have been created to implement ideas borne of science and engineering [. . .] It exists to manage and modify the physical and biological world in a constructive way and relies on a foundation of mathematics [. . .] Technology systems result from engineering design and development.' (*NSF, 1983b, p. 73*)

Occasionally there are slips from the kind of view of technology presented above, which reinforce the feeling that science is the superior 'partner'. For example, 'science is best introduced as "applied science", which is technology'. However, despite this, the recommendations give a rounded view in terms of the technological systems and concepts:

Technological systems	Concepts
Communications	Problem formulation and solving
Energy production and conservation	Debugging a problem
Transportation	Discovering alternative solutions to problems
Shelter	Making connections between the parts, theory and
Food production	practice
Health care delivery	Pattern recognition
Safety	Engineering approaches to problems; evaluation of
Residential use of space	trade-offs
Resource management	Probability/approximations/estimation
Biotechnology	Building and testing equipment
Computers	Examining trade-offs and risk analysis
Nuclear issue	Economic decision-making
	Feedback and stability
	Recognizing orders of magnitude

In fact most of the so-called 'concepts' are actually processes, and the technological systems sometimes are issues (e.g. nuclear) or topics (e.g. shelter).

The term used for technology education is *technological literacy*, defined as the 'understanding of behaviour of technological systems [requiring] a knowledge of scientific and mathematics concepts and those unique to engineering' (NSF, 1983b, p. 74). The inconsistencies in the curriculum proposals over the view of technology education occur, for example, in not combining the problem-solving objectives with the ideas on design and synthesis, such that the study of case studies in design is suggested, rather than doing any design. But, despite this, there is an extremely sophisticated view of problem solving presented in the source material, that presents a discussion of what is known about learning technology. For example, the importance of qualitative understanding (of, say, how an electronic system works) being necessary before beginning any quantitative work, and the importance of knowledge in solving problems (see Article 3.1).

■ Project 2061

Like the previous project, this one was based upon the recognition that young Americans were performing poorly on international tests. But it also recognized the need to prepare them for a changing world of rapid growth in scientific knowledge and technological power. The project was named *2061* because when it started in 1985 Halley's Comet was overhead and in the year 2061 it will next be overhead. This indicates the timescale envisaged; the first phase, reported in 1989, was to establish a conceptual base for the reform of knowledge, skills and attitudes to be acquired by students (AAAS, 1989a and b; Johnson, 1989).

The fact that this was an American Association for the Advancement of Science (AAAS) project, and indeed was also called *Science for All Americans*, accounts for a bias towards science. As with the NSF project, however, there was a real appreciation of the uniqueness of technology. Similar conflicting messages are evident. For example, on the one hand a separate Technology Panel was set up, but to address the basic question 'What is the technology component of scientific literacy?' (Johnson, 1989, p. viii). Nevertheless the reports show a recognition of unique technological principles (e.g. systems and control), and the distinction between the 'designed world', which is the business of technology, and the natural world for science. But the dependence of technology on scientific and mathematical understanding is continually stressed. Indeed there are a number of common themes: systems, models, stability, patterns of change, evolution and scale (AAAS, 1989b, Chapter 11). Some of these would of course be seen as central features of technology, but the project seeks to explore how they are shared by science, mathematics and technology.

Another stress is upon the value issues in technology, which is evident through the recognition that technology is part of the culture system and both shapes and reflects the system's values (AAAS, 1989b, p. 39). It is also evident in the 'principles of technology' (AAAS, 1989b):

(1) The essence of engineering is design under constraint.

(2) All technologies involve control (e.g. using feedback).

(3) Technologies always have side-effects.

(4) All technological systems can fail.

Part of the preparation for young people to cope with the changing world (noted earlier) finds expression in a concern about the relationship of technology and society. All citizens should know the relevance of questions like: who benefits, what are the costs of new technology, what are the risks?

The Technology Panel report sets out to identify the concepts and skills of most significance for technology: not based upon the current curriculum, and showing awareness of changes in technology. The panel defined technology as:

'. . . the application of knowledge, tools and skills to solve practical problems and extend human capabilities.' (*Johnson, 1989, p. 1*)

But the panel also added that it 'is best described as a process, though more commonly known by its products and effects on society' (Johnson, 1989, p. 1). Although there is a recognition of general technology 'concepts' such as *work*, *flow*, *design*, *innovation*, and *risk/benefit*, these are not spelt out in great detail. What is more clearly delineated are the 'technologies' that students should encounter in their technology education: materials, energy, manufacturing, agriculture and food, biotechnology and medical technology, environment (atmosphere), communications, electronics, computer technology, transportation, and space. These represent a wider range than those of the Jackson's Mill analysis.

There is also an extensive discussion of technology education, including what should be the nature of courses, the role of IT, the importance of the use of science and mathematics, the learning of concepts, and what has to be learnt about the interface of technology and society. Several aspects are picked as important.

- *Problem solving*, recognizing that there are different kinds: technical, experimental, mathematical, technical–social, or value-laden.
- *Design alternatives* that require students to learn to deal with options.
- *Observation*, *measurement* and *analysis* as universal tools of technology.
- *Intelligent observation* crucial to *invention*, requiring *creative thinking*.
- The understanding of patents and copyrights.
- The need for *imagination*.
- The need to *question* basic assumptions, while seeking solutions, learning to *visualize the whole* while looking at the parts.
- Distinguishing *possibilities* from *probabilities* and decision-making with incomplete information.
- Ability to *collaborate* and *cooperate*.
- *Obtaining* and *organizing* information.
- Having a *strategy for learning* what is relevant and what is not.

For each of the technologies (materials, energy, etc.) the importance of the technology and what it includes are outlined, along with the kind of student experiences that are envisaged. For *materials* the following is said:

'Traditional industrial arts programs are named after materials (such as wood or metals) or their processing (such as foundry or welding). Students should have some ongoing experiences with making artifacts of wood and metal during their school years, as well as with other traditional hands-on activities such as typing (currently, keyboarding or data entry), cooking and sewing. But hands-on practice with regard to materials would be extended to include some of the advanced materials used in electronics, composite materials, and biomaterials. A central focus for these experiences should be determining properties, first qualitatively (in the fifth grade, for example, making structures of soda straws or toothpicks to show compressive or tensile strength) and later quantitatively (in high school, for example, using testing equipment to make numerical measurements of the strength of various materials).' (*Johnson, 1989, p. 14*)

The strong desire to develop science is also evident in the suggestions for *agriculture and food*:

'Curricula should emphasize the techniques of agricultural technology, including advanced biotechnology as it becomes available at the elementary and secondary school level. Students should be involved in experiments with plants, animals, insects, fungi, molds and other life forms. They should also have direct experience with the effects of fertilizers, animal nutrition, plant physiology and ecology. The implications of the laboratory results should be discussed in concurrent social studies classes to continue the learning process. Many schools have laboratory facilities for food preparation. Those activities should be encouraged, and extended to include experiments related to nutrition and home, industrial and commercial food processing and packaging.' (*Johnson, 1989, pp. 18–19*)

For *communications* the experiences are more 'design and make':

'Students can make simple devices that are used in communications, from historical gadgets (such as a carbon microphone or a simple telegraph) to modern electronic circuits, and they can then use them in elementary networks of their own design. Students should be encouraged to undertake imaginative projects, such as inventing ways of communicating with people in remote lands or searching for information from outer space that might reveal life there.' (*Johnson, 1989, p. 23*)

The second and third phases of this project will try out some of these ideas in schools to develop specific curricula, and to work with scientific societies and education institutions to reform technology education.

■ A conceptual framework for technology education

This was produced through the auspices of the International Technology Education Association (ITEA), which represents the interests of technology educators mainly in North America (ITEA developed out of the association for industrial arts). In a sense this was an attempt to build upon the ideas of Jackson's Mill, cast in the mould of 'doing' technology, and hence with a central idea of problem solving geared to serving human needs and wants. The conceptual framework is built around a 'technological method model' which has at its heart problem solving. Indeed, the 'primary method of knowing for technologists is defined as a problem-centred approach, governed by human needs and wants' (Savage and Sterry, 1990, p. 12). This view of technology leads to a focus on 'doing', emphasizing process education through problem solving.

This view of technology ascribes a number of attributes to it:

- People create technology.
- Technology responds to human wants and needs.
- People use technology.
- Technology involves actions to extend human potential.
- The application of technology involves creating, implementing, assessing, and managing.
- Technology is implemented through the interaction of resources and systems.
- Technology exists in a social/cultural setting.
- Technology affects and is affected by the environment.
- Technology affects and is affected by people, society and culture.
- Technology shapes and is shaped by values. (*Savage and Sterry, 1989, p. 11*)

Technological method is treated as being parallel to scientific method. It envisages *problem solving* drawing upon *resources*, and *technological processes*, to deal with problems or opportunities that result in outcomes and consequences. I will explain the terms in italics.

Problem solving is seen to have six steps:

- Defining the problem.
- Developing alternative solutions.
- Selecting a solution.
- Implementing and evaluating the solution.
- Redesigning the solution.
- Interpreting the solution. (*Savage and Sterry, 1990, p. 14*)

Resources include people, tools/machines, information, materials, energy, capital and time.

Technological processes are in fact areas of content, and four kinds are identified: bio-related technology, communication technology, production technology, and transportation technology. These appear on the face of it to be the same as the Jackson's Mill human adaptive systems, except that they are defined in terms of *process techniques*. Thus for bio-related technology these techniques are: propagating, growing, maintaining, harvesting, adapting, treating and converting.

Technological education is defined as the *study* of technology and its effect on individuals, society and civilization, with the mission of preparing individuals to comprehend and contribute to a technologically-based society. The use of the word 'study' is actually a distraction because the goals are to achieve technological literacy and capability, defined as:

(1)   Utilizing technology to solve problems or meet opportunities to satisfy human needs and wants. [. . .]

(5)   Evaluating technological ventures according to their positive and negative, planned and unplanned, and immediate and delayed consequences. (*Savage and Sterry, 1990, p. 20*)

The design and creation of objects is noted as one of the activities of relevance to technology education, and it is the only kind of 'doing' that is specifically identified. However, the six steps of problem solving do not imply only design and making as the activity.

This framework is intended to allow a teacher to plan a programme, although not all elements (e.g. technological processes) are spelt out in sufficient detail. The focus on problem solving, if taken up by all technology teachers, will have a substantial impact on technology education. Time will tell if such proposals have this effect.

■ An overview

These three proposals show some similarities and some differences. They all show a concern to distinguish technology from science, although the first two are not always consistent in this. They all see an important role for problem solving. Design is recognized as a part of the technological activity, but does not dominate, and is certainly not a central part of student experience. It may be that this will occur in practice when the ITEA proposals are implemented, but this is not defined as central to it.

All of the proposals are strong on defining the content of technology, and here they reflect the tradition in the USA. Compared to the UK their definitions of content are more extensive, and in some ways more ambitious. Whether they can be experienced in an active way by students as is hoped is a different matter. But they offer an interestingly different view of technology as part of the culture than in the UK.

Most of the proposals (ITEA's less so) show a strong concern for value issues and preparing future citizens for a technological society. As noted earlier this to some extent illustrates the influence of the STS view of technology. However, in the first two proposals the STS view is taken further to include more active doing of technology, not just studying it.

As with all curriculum proposals, what is contained in documents will only take effect when teachers share the understandings they contain. Implementation in schools is quite a different issue from proposing, as Article 5.6 illustrates.

■ References

American Association for the Advancement of Science (AAAS) (1989a) *Science for All Americans: Summary. Project 2061*. Washington, DC: AAAS.

American Association for the Advancement of Science (AAAS) (1989b) *Science for All Americans. A Project 2061 Report on Literacy Goals in Science, Mathematics and Technology*. Washington, DC: AAAS.

Booth, B. (1990) The development of technology education in the United States. *Studies in Design, Education, Craft and Technology*, 21 (2), 84–9.

Eyestone, J.E. (1989) *The Influence of Swedish Sloyd and its Interpreters on American Art Education*. Unpublished PhD. thesis. Columbia: University of Missouri.

Hacker, M. and Barden, R. (1983) Systems approach to technology education. *Man Society Technology*, 42 (6), 9–14.

Johnson, J.R. (1989) *Technology: Report of the Project 2061 Phase I Technology Panel*. Washington, DC: AAAS.

Luetkemeyer, J.F. (1985) The Social Settlement Movement and Industrial Arts Education. *Journal of Epsilon Pi Tau*, 11 (1–2), 97–103.

National Science Foundation (NSF) (1983a) *Educating Americans for the 21st Century*. Washington, DC: NSF.

National Science Foundation (NSF) (1983b) *Educating Americans for the 21st Century. Source Material*. Washington, DC: NSF.

Savage, E. and Sterry, L. (eds) (1990) *A Conceptual Framework for Technology Education*. Reston, VA: International Technology Education Association.

Snyder, L. and Hales, J. (1981) *Jackson's Mill Industrial Arts Curriculum Theory*. Charleston: West Virginia Department of Education.

PART 2

Technological Capability and Practical Action

2.1

Technological Capability

P. Black and G. Harrison

[It was in 1985 that Professors Paul Black and Geoffrey Harrison produced their pamphlet 'In place of confusion' that contained an exposition of the nature of technological capability. In this extract from that pamphlet, they translate general ideas on capability into specific activities in the classroom aimed at developing technological capability.]

■ Introduction

If we want to know how we should educate children in and through technology, we must first answer two questions:

(1)   What is technology?
(2)   For what purposes should it play a part in children's education?

This article attempts to answer these questions. The argument is developed as follows:

(a)   Section 1 attempts to define the essence of technology.
(b)   Section 2 looks at human capability in more general terms.
(c)   Section 3 sums up the argument with the concept of task–action–capability.
(d)   Section 4 gives examples of situations in which such capability can develop.

■ The essence of technology

Technology is the practical method which has enabled us to raise ourselves above the animals and to create not only our habitats, our food supply, our

comfort and our means of health, travel and communication, but also our arts —
painting, sculpture, music and literature. These are the results of human
capability for action. They do not come about by mere academic study, wishful
thinking or speculation. Technology has always been called upon when practical
solutions to problems have been called for. Technology is thus an essential part
of human culture because it is concerned with the achievement of a wide range
of human purposes.

In the mid 1960s, those wanting technology to play an important part in
education asked 'What is technology?' The question was answered in the
following way. 'Technology is a disciplined process using resources of materials,
energy and natural phenomena to achieve human purposes.' This definition led
to three complementary sets of educational aims:

(1) To give children an *awareness* of technology and its implications as a
     resource for the achievement of human purpose, and of its dependence on
     human involvement in judgemental issues.

(2) To develop in children, through personal experience, the *practical
     capability* to engage in technological activities.

(3) To help children acquire the *resources* of knowledge and intellectual and
     physical skills which need to be called upon when carrying out
     technological activities.

However, the above definition and aims have not been totally accepted or
understood. Some teachers have concentrated their effort on practical capability,
to the neglect of other aspects. Others have emphasized the resources and given
little attention to their use. Emphasis on its many harmful effects has called in
question the value-free promotion of technology — thus exposing problems
about aim 1. Such difficulties suggest that the definition and the aims need to
be re-examined.

So, what are the questions which *should* be asked?

How do we describe, and educate our children for, those human activities
which bring about change, enhance the environment, create wealth, produce
food and entertainment, and generally get things done? What is the nature of
capability in these activities, how can it be fostered and what kind of back-up
knowledge and experience is needed? How can future citizens be better
equipped to foresee consequences and make choices?

■ Human capability

First, let us consider a range of such activities which, although diverse, do
perhaps have a common pattern. Then let us examine the implications for

education and its opportunities and responsibilities for fostering such capability in young people.

Human capability lies at the heart of such diverse activities as:

- creating a self-propelled flying machine
- composing a symphony
- writing and directing a television show
- organizing an office business system
- managing a mixed arable and livestock farm
- creating a three-dimensional mural for a public building.

The activities need not be on a grand scale. Capability is also called for in:

- designing and building a garden shed
- writing and producing a sketch in a school revue
- setting up a system of domestic accounts
- maintaining a car or bicycle
- putting up shelves
- carving a piece of sculpture
- hanging wallpaper.

Large or small, these activities call for a variety of competencies which the capable person knits together in order to achieve success. Maintaining a motor car requires competence in mechanical and electrical fault-finding, in correct and skilful use of tools, in treatment and preservation of materials susceptible to corrosion, in manipulating heavy equipment with safety.

Similarly, setting lyrics to music calls for imagination and intuitive flair. It also calls for perception of meaning in words and in music and an ability to match one to the other. In addition to these imaginative and creative processes, the composer needs to have at his or her fingertips an understanding of harmony, melody, rhythm and structure.

A similar analysis could be made for all of the examples. They have a common pattern. Each requires:

- application of personal driving qualities such as determination, enterprise, resourcefulness
- personal innovative powers of imagination, intuition and invention
- powers of observation and perception
- willingness to make decisions based both on logic and on intuition
- sensitivity to the needs being served, to the possible consequences, benign or harmful, of alternative solutions, to the values being pursued.

However, overlapping all these is the common necessity to possess a sound base of knowledge and both intellectual and physical skill appropriate to the job in hand. The shed builder must know about the treatment and processing of timber, about how to make effective connections between structural members and about principles of strength and weakness, rigidity and stability, weatherproofing and foundations. The composer needs to know the principles of harmony and rhythm and have the skill to perform on musical instruments. The office manager needs to know the principles of accounts and the techniques for management relevant to his or her business. The farmer needs to know about fertilizers, pesticides, basic medical treatment and the technical requirements of machinery, plant, equipment and building.

Thus the common pattern shows that full capability for personal action calls simultaneously for both action-based qualities and the resources of knowledge, skill and experience.

The first without the second may lead to frustrated, hyperactive, but ineffective individuals. The second without the first leads to individuals who are highly knowledgeable and skilled but who may be incapable of producing new solutions to problems.

This interaction between the *processes* of innovative activity and the *resources* being called upon is itself one of the key elements of successful human capability. It is a continuous engagement and negotiation between ideas and facts, guesswork and logic, judgements and concepts, determination and skill.

■ Task–action–capability

If the nature of these personal human attributes which bring about a capability to engage in active tasks is becoming clearer, the second question remains to be answered. How do we educate our children with a view to maximizing their individual potential for what might be called 'task–action–capability' (TAC)?

There are three dimensions to TAC which are amenable to educational development. Each might be considered central from particular and different points of view, but, nevertheless, each represents a personal attribute of direct practical value in the real world:

(1) *Resources* of knowledge, skill and experience which can be drawn upon, consciously or subconsciously, when involved in active tasks.

(2) *Capability* to perform, to originate, to get things done, to make and stand by decisions.

(3) *Awareness*, perception and understanding needed for making balanced and effective value judgements.

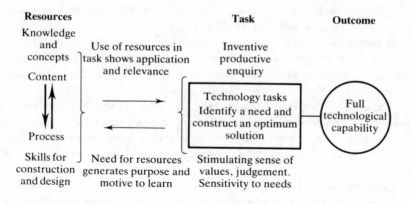

Figure 1 A model of technology education.

These three clearly interact. To develop capability and awareness, experience of tackling tasks is essential. Through such tasks we learn how to use and apply resources of knowledge and skill, for the mere possession of such resources does not imply or confer the ability to apply them. The relationship is mutual, for the needs of real tasks can provide a motive for acquiring new knowledge and skills or for consolidating those already learnt.

This mutual interaction between resources, and tasks chosen to develop capability and awareness is represented in Figure 1.

■ Tasks for learning

In the particular areas of engineering and design, where the concepts of science and technology play essential roles, the three dimensions of TAC become very obvious. Nevertheless, in order to be able to construct learning systems which will be effective in the overall development of TAC all three dimensions will need to be planned, interwoven and modulated to meet the needs which change with age, ability, interest and motivation. Three brief examples may help to illustrate this point.

☐   Task 1   Moving loads up a long ramp

Thirteen-year-old children were given this task to help develop learning about technological concepts of energy (a *resource*).

They were given assorted motors, electricity supplies, gears, pulleys, wire and other raw material. By giving different briefs to different groups, recognizing diverse abilities, the children were faced with appropriately challenging opportunities to engage in inventive problem solving (and so develop their process skills as another *resource*).

Between them they were expected to identify, use and compare three choices in transport technologies: the use of a locomotive with a self-contained power unit: electrical power delivered to a vehicle along wires; and power delivered to a vehicle mechanically by string.

The need to compare these led to the idea of power/weight ratios which in turn led to structural design concepts for the vehicles themselves. At this age the concepts could not be quantified except in general comparative terms. It was the beginning, however, of the development of all three dimensions of TAC within the one project.

Children learned to develop *resources* – problem solving, skills of observation, experiment, evaluation, designing and decision-making, and concepts of power and energy.

Children learned to develop *practical capability* in using these resources. The varied approach led to varied solutions bringing with them the understanding that there is no single correct solution; one had to be evaluated against another, for which criteria needed to be defined. Children learnt that, from the smallest detail (say, wheel bearings) to the overall concept, decisions have to be taken in order to achieve success.

This exercise also promoted wider discussion of examples from all over the world: the two common forms of electric train in this country; the diesel-electric locomotive, diesel multiple units; the cable cars of San Francisco; trams and trolley buses in continental cities. So children developed an *awareness* that the real world has no single optimum solution. The factors which influenced decisions were seen to be environmental; economic costs, energy resource implications, and relations between these and technological optimization could all be touched on in order to awaken a concern for *value judgements*, including moral and aesthetic, in technological developments.

☐   **Task 2   The hybrid car**

This task involved a class of 15-year-olds in an inner city school which, as a group, designed the complete system for a car propelled by a hybrid of petrol and electrical propulsion. The design process involved children in acquiring new *intellectual resources*, including detailed knowledge about high current electronic control circuitry, and about the mechanical, structural and dynamic principles essential in a road vehicle which has to conform to the Road Traffic Acts and win awards in the BP Buildacar competition.

The main purpose of this project was to motivate a group of children to become fully involved in a real task which would not only develop their inventive and design skills and their manufacturing capability but do two further things. It helped those children to become *aware* of the importance of learning some science in order to achieve something useful; it also brought about a vivid realization of the *value judgements* involved when conservation of energy resources and the potential impact of the internal combustion engine on the environment had to be considered.

☐ **Task 3   Building the motorway**

Both of the previous examples involved children in designing and making as a process. The tasks that might be set for such processes can be modulated to take account of the maturing minds and skills. They can be appropriately modelled for the youngest ages in the primary schools. However, studies may also be more investigative than design-based, and these too can be adapted to all ages.

For instance, a group of primary school children were engaged in a topic that focused on the construction of a nearby motorway. They had to engage in various *processes* of enquiry: they visited the site and talked to construction workers, nearby residents, and a farmer whose land was being used; the discussion of the advantages and disadvantages of the change raised the problems of choice and conflicting *values*. They went on to study methods of road construction and examined different soil samples, so acquiring new *resources* of knowledge. They also built models of soil-moving machinery using electric motors, simple levers and gears. The theme also involved them in map reading, in studying transport past and present, in drawing and in writing poetry.

The list of possible examples is endless, but they should not be seen as isolated examples. Any successful example illustrates kinds of capability appropriate to particular stages in the development of the individuals involved. An essential condition for success is that the tasks be structured progressively, comprehensively and in close co-ordination between those areas of the curriculum which can contribute to resources of knowledge and skill and those which can help in making the judgements which the exercise of modern technology forces on society.

Indeed, it is only when the three dimensions of TAC have been properly developed that young people become able to take part in decisions, whether these be the complex decisions of any democratic society – such as those concerned with transport, land use and services – or the decisions which face individuals seeking to make a go of running their own homes and gardens.

Learning through Design and Technology

Assessment of Performance Unit

[This is the first of two extracts in this Reader from the report by the APU of their extensive investigation into the nature of design and technological capability and its assessment. This one reviews different models that have been used to describe design and technology and outlines the one developed by the research team. It explores the dimensions of this model and the way these can be used to form the basis of an assessment programme for design and technology.]

■ Introduction

[. . .]

From the earliest work in this field, there has been general agreement on certain basic tenets of design and technology. It is an *active* study, involving the *purposeful* pursuit of a *task* to some form of *resolution* that results in *improvement* (for someone) in the made world. It is a study that is essentially procedural (i.e. deploying processes/activities in pursuit of a task) and which uses knowledge and skills as a resource for action rather than regarding them as ends in themselves. The underlying drive behind the activity is one of improving some aspects of the made world, which starts when we see an opportunity to intervene and create something new or something better.

[. . .]

■ Models of design and technology

If the motives underpinning the activity have been the source of some contention, things get even more difficult when people attempt to describe exactly what the process consists of. Early models of the activity described it in simple problem-solving terms that start with a problem and progress through a linear sequence of steps to a solution [Figure 1].

Innumerable variants of this basic idea can be found in the literature and gradually – as teachers became more experienced at working with them – the models were themselves refined. The linear track became a 'design loop' on the reasonable grounds that the evaluation of the end product must not only be conducted in relation to the initiating problem, but that, moreover, the results of the evaluation will themselves provide new problems to start the cycle all over again [Figure 2].

Once again, however, familiarity with this process led to dissatisfaction with it as an adequate description of what pupils do when they tackle a task. Do they not have to look things up (research them) when they are making things, or generate ideas (e.g. about testing methods) to evaluate a solution. And do they not have to evaluate ideas as they emerge to see if they are worth pursuing? The models used to describe the process became ever more confused, as the subtlety of the process became apparent.

The principal motives behind this drive to analyse the constituent parts of the activity lay in the need to make it possible to *teach* and *assess* it. For these two purposes it became increasingly necessary to try to impose order on what is essentially a confused, interactive process. These models have been helpful guides to the sorts of activity that need to go on in design and technology, but they have equally been dangerous in prescribing 'stages' of the process that need to be 'done' by pupils.

> 'Used unsympathetically, the approach can reveal a greater concern for "doing" all the stages of the process, than for combining a growing range of capabilities in a way which reflects individual creativity and confident and effective working methods.' (APU, 1987, p. 2.12)

The essence of this problem lies in the transformation of active capabilities into passive products. To take an example, 'investigation' or 'research' is typically one of the stages identified in the process and results in the

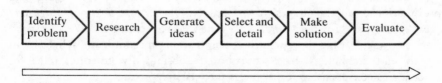

Figure 1   A simple linear model.

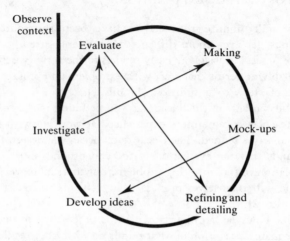

Figure 2   An interacting design loop (source: Kimbell, R./SEC/OU 1986 OU Press).

accumulation of large folders of background material related to the task. In making an assessment of this 'stage' of the process, where should we look for evidence of a pupil's investigative/research capability? Naturally we look at the folder. Inevitably, therefore, the research folder (a product) comes to represent an active capability. There are three related problems here.

First, the development of the investigation folder typically assumes that all investigation goes on early in the project, i.e. it is about investigating the *task* to see what is involved in tackling it. But do we not need constantly to be investigative? Don't we need to investigate user reactions to our early design ideas, or the most appropriate glue for a particular making task? Typically, by allocating 'investigation' to a particular stage of the process, and by committing it to the formality of a separate folder, we prevent pupils from recognizing the need *to be investigative* at all times.

Second, because the investigation *folder* becomes the focus of assessment there is an enormous pressure to 'pretty-up' the folder after the event, regardless of its relevance to the developmental process and as if it were to be valued as a product in its own right. Which frequently of course it is.

Third, this line of reasoning results in the efficient packaging and presentation of all the 'stages' of the process. *In fact, the process of design and technology becomes a series of products* (*The* Brief. *The* Specification. *The* Investigation, etc., etc.). It has to be said that examination procedures and syllabuses in design and technology have contributed substantially to this unhelpful tendency to convert active capabilities into passive products.

While the analysis of the process into these discrete elements may have helped teachers to get to grips with the parameters of what is involved in tackling a design and technology task, it has frequently emasculated it by ripping it apart in quite unnatural and unnecessary ways. Assessment in design and technology has too often assumed that you can measure the quality of an omelette simply by measuring (and aggregating) the individual quality of the eggs, the milk, the butter and the herbs. As if the cook was irrelevant!

■ The interaction of mind and hand

For [the] APU we attempted to create a different way of looking at design and technology; a way that placed the interactive process at the heart of our work and the products as subservient to that process. To do this, we rejected the idea of describing the activity in terms of the products that result from it, and instead concentrated on the thinking and decision-making processes that result in these products. We were more interested in *why* and *how* pupils chose to do things than in *what* it was they chose to do. The pupil's thoughts and intentions were as important to us as were the products that resulted from them.

We gradually came to see the essence of design and technology as being the interaction of mind and hand – inside and outside the head [Figure 3]. It

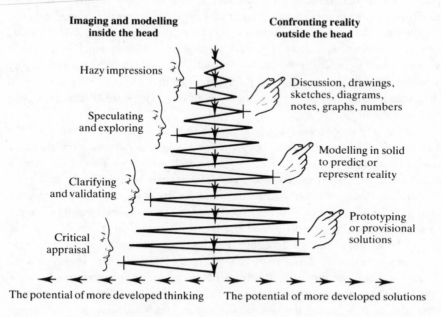

Figure 3 The APU model of interaction between mind and hand. [The authors] refer to this model throughout [. . .] as a stylized triangle. Wherever this appears it is intended to convey the detail of the full version.

involves *more* than conceptual understanding – but is dependent upon it, and it involves *more* than practical skill – but again is dependent upon it. In design and technology, ideas conceived in the mind need to be expressed in concrete form before they can be examined to see how useful they are.

☐   **Imaging in the mind**

It is not uncommon for pupils to believe that, almost from the start of an activity, they have a complete solution sorted out in their mind and this often leads them to try to short-circuit the process of development. 'I know what I want to do – I just need a piece of plywood/felt/clay . . .' In fact we know that they cannot, in their mind alone, have sorted out all the issues and difficulties in the task – let alone reconciled them into a successful solution. What they have got is a hazy notion in their mind's eye of what a solution is like – and this is a crucial starting point for them. But it is only a starting point, and to enable the idea to develop it is necessary to drag it out of the mind and express it in real form. To demonstrate this phenomenon, simply imagine an 8-year-old girl with spinal injuries lying in a hospital bed. She has to lie flat on her back and not move. She loves doing jig-saw puzzles. Can we develop for her a jig-saw puzzle system that can operate effectively within these constraints? Even with this tiny amount of information we have all begun to image – in our mind's eye – solutions that we believe might work. The solution might involve magnets, or velcro, or possibly a sheet of glass or clear plastic. This internal image is our starting point from which we may – eventually – be able to fashion a satisfactory solution. But it is only a starting point, and our first responsibility is to try to drag this internal image out into the light of day. There are at least two good reasons for this.

First, the process of trying to express a hazy idea forces us to clarify it. We soon see that our vague idea about magnets might mean two very different things. Are we to stick little magnets onto each piece – or should we make a new puzzle out of magnetic material? The former sounds fiddly and time consuming – but the latter would eliminate the use of all existing jig-saws. By trying to express our idea – in words, or pictures, or in concrete reality – we get closer to seeing the difficulties and the possibilities within it.

The second reason for dragging the idea out of our mind and expressing it in some way is that by doing so we make it possible for others to share our idea. As soon as your idea is expressed – in words, or pictures, or in concrete reality – it is something that I can comment on. 'Do you mean . . .' 'Do you really think that . . .' 'But what if . . .' As teachers, this is one of our major responsibilities, to act as a catalyst by providing helpful, critical but supportive comments on pupils' developing ideas.

For these two reasons, therefore, the act of expression is a crucial part of the development of thinking. Without such expression it is almost impossible for an idea to move very far forward because very few people are able to cope with that degree of mental imaging. It is like playing mental chess. We can all manage the first move or two – but trying to hold in our mind an image of the board after 20 moves (and counter-moves) is impossible for most of us. With the chess board in front of us (as a concrete expression of the current state of our thinking) we can achieve a far more cunning and sophisticated level of thinking. So too with design ideas – the concrete expression of them not only clarifies them for us, but moreover it enables us to confront the details and consequences of the ideas in ways that are simply not possible with internal images. Cognitive modelling by itself – manipulating ideas purely in the mind's eye – has severe limitations when it comes to complex ideas or patterns. It is through externalized modelling techniques that such complex ideas can be expressed and clarified, thus supporting the next stage of cognitive modelling.

It is our contention that this interrelationship between modelling ideas in the mind, and modelling ideas in reality is the cornerstone of capability in design and technology. It is best described as 'thought in action'.

☐   **Modelling solutions**

Expressing ideas is therefore a necessary part of developing ideas. But does it matter what mode of expression is used? Are all means of expression equally applicable to any design circumstances? Clearly not, and part of the art of developing capability in design and technology is to develop a rich variety of modelling strategies that enable ideas to be expressed in the most appropriate ways.

Choosing the most appropriate form of modelling involves thinking not only about what the idea is that needs to be expressed, but equally about how the modelling is supposed to help. If we wish to explore a basic concept for a new product (e.g. a new car radio volume control that adjusts the volume according to the ambient noise level) then, at this level of broad concept, *discussion* (verbal modelling) may be the best way to start. It is very quick and it helps people to get a grip on some of the big issues and difficulties that might need to be tackled. It is interesting to note that in the modelling tests that we developed, discussion proved to be a vital element in pupils' responses. For some situations they found it to be the most useful form of modelling.

But discussion alone does not allow us to get into the detail that would be required to evaluate how the noise sensing could best be achieved, let alone confront the details of electronic circuit design. Different types of modelling are needed that may be *diagrammatic*, or *computer-simulated* and that enable fine detail to be explored and resolved. As the electronics is being sorted out it will probably be necessary simultaneously to consider the styling and ergonomics of the developing product, and this needs a different form of modelling again,

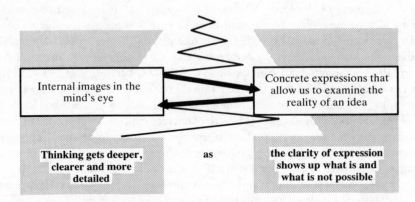

Figure 4   Clearer and more detailed expression allows clearer and more detailed thinking.

probably involving a range of *graphic techniques* and *3D models* that fully represent the appearance and feel of the finished article.

With modelling it is a matter of horses for courses. Is speed the priority or accuracy; is the idea qualitative or quantitative; are you trying to sort out the visual details or the mechanical functioning; analyse the stresses in a shelf support or detail the combination of flavours and spices in a new snack product? There are many ways of modelling ideas and each has its advantages and disadvantages. Accordingly pupils need a rich awareness of the diversity of possibilities for modelling to enable them to grapple with the particular requirements of their task [Figure 4].

But the guiding principle behind the selection of any technique must be that it enables pupils progressively to confront the reality of their ideas – to get a better and sharper grip on what they will be like (and how they will work) when they are finally completed. We must recognize, however, that the notion of the finally completed product is elusive for there is really no such thing. 'The end' of the process is typically arbitrary and determined not by the task itself so much as by the parameters that bear on it – typically the timescale of the project. There is nothing particularly special about the 'final' prototype, for the moment it exists it becomes the focus for yet further refinement and is therefore but another extension of modelling activity. In the commercial world, the endless progression of updated versions of existing products bears witness to the possibilities for refining and developing ideas. While the motive underlying these developments will probably be the maintenance of market share in a fickle consumer world, the process of development has modelling at its heart. If scientific innovation (micro-chips, lasers, etc.) often provides the opportunity for product development, it is modelling that provides the dynamic driving force that carries development forward from hazy ideas to refined and detailed working prototypes.

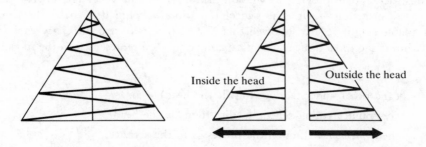

Inside the head

Outside the head

Figure 5   Traditional assessment strategies split the interactive process and use different assessment techniques.

This active, task-centred description of design and technology carries with it a number of implications, the most central of which concerns the dimensions of an assessment framework for design and technology [Figure 5].

■ The dimensions of capability

The model presents us with two completely contrasted ways of looking at the assessment of capability. Traditionally, assessment devices in design and technology have separated the conceptual and the expressive; conceptual matters being seen as testable in written examinations, while the expressive are equally (though perhaps less validly) testable in graphic/modelling/making examinations.

We took the view – supported by the brief we had been given – that this approach to assessment is destructive of the essence of capability as we have described it above. We were not interested in conceptual understanding for itself, or in the decontextualized display of any particular communication skill but rather in the extent to which pupils can *use* their understandings and skills when they are tackling a real task. Capability in design and technology involves the active, *purposeful deployment* of understanding and skills – not just their passive demonstration. Isolated tests of knowledge and skills were therefore quite inappropriate and we had to look towards the development of test tasks that could give us a measure of active capability.

This idea, when applied to our model of the activity, meant that we had to consider a completely different way of looking at tests. If the integrity of the imaging and modelling (inside/outside, conceptual/expressive) process were to be maintained, then any splitting up of the process had to be done without destroying this crucial relationship.

Given this starting point, we developed the idea that tests might be constructed that provide a 'window' through which we could observe the process in action – with the size of the window being defined by the time available [Figure 6]. It followed that any such test must of necessity contain within it the three dimensions of capability that are represented in the model, i.e.

- conceptual understanding (inside the head)
- communicative/modelling facility (outside the head)
- their *interaction* through the processes in the activity.

By slicing the process in this way we hoped that it would be possible to see (and assess) the central procedures of the activity as well as the extent to which they were resourced by conceptual understanding on the one hand and expressive facility on the other. We thus derived the three principal dimensions of an assessment framework.

More than simply identifying these dimensions however, the model has further implications. It is as important for pupils to be aware of what they *need to know* as it is for them to actually know it. Accessing new knowledge and skills in response to the demands in the task is a fundamental characteristic of capability in design and technology.

'The designer does not need to know all about everything so much as to know what to find out, what form the knowledge should take, and what depth of knowledge is required for a particular purpose.' (*DES/APU, 1981, p. 5*)

It therefore became as important for us to probe a pupil's capability in accessing relevant knowledge and skill as it was to register the existing knowledge that they were already using in their task.

[. . .]

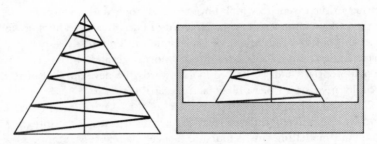

Figure 6 Looking at tests as windows, through which to scrutinize parts of the process while maintaining the integrity of the model.

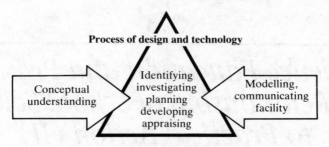

Figure 7 The dimensions of capability.

■ References

Assessment of Performance Unit (APU) (1987) *Design and Technological Activity: A Framework for Assessment*. London: APU/DES.

DES/APU (1981) *Understanding Design and Technology*. London: DES/APU.

Kimbell, R. (1986) *Craft Design & Technology*. Milton Keynes: The Open University Press.

2.3

Science Education and Praxis: The Relationship of School Science to Practical Action (II)

D. Layton

[This is a section of an original article by Professor David Layton, an earlier section of which appears as Article 1.2 in Reader 1 *Technology for Technology Education*. In this section, Layton considers the implications for pedagogy, institutional structures, and research, of his views about knowledge and practical action.]

[. . .]

■ Other approaches to cognition in practice

Although debate about the nature of practical knowledge goes back at least to Aristotle (Jonsen and Toulmin, 1988, pp. 58–64), it has not been a preoccupation of modern philosophers. In one of the few recent books which attempts a direct attack upon the question, Barry Smith acknowledges that, over 40 years after Gilbert Ryle's paper on 'Knowing how and knowing that' in 1945, 'the problem of practical knowledge has still failed to establish for itself a secure position in the field of problems dealt with by analytical philosophers' (Nyiri and Smith, 1988, p. 1).

Signs that the situation is changing are discernible, though not yet plentiful. One example is Antonio Pérez-Ramos's study of 'the maker's knowledge tradition', with its exploration of 'reason as portrayed in human purposive action (doing/making), especially when based on the natural sciences' (Pérez-Ramos, 1988, p. 3). In a critique of Hirstian forms of knowledge and in the context of deliberations about school technology, Neil Bolton has drawn upon the ideas of Michael Polanyi and Maurice Merleau-Ponty to argue that 'all knowing depends upon the practical as a means of developing personal commitment' and that, in the development of any subject area, 'the practical is both foundational and the continuing source of new knowledge' (Bolton, 1987, pp. 10–11). It is in the field of moral philosophy, rather than that of epistemology, that philosophers have shown greater interest in problems of practical knowledge and practical reasoning, however. Recent advances in the biological sciences, and in genetic manipulation especially, have engaged their attention, with works such as Peter Singer's *Embryo Experimentation* (1990), *In Defense of Animals* (1985) and *Applied Ethics* (1986) being examples from a burgeoning genre. Albert Jonsen's and Stephen Toulmin's *The Abuse of Casuistry. A History of Moral Reasoning* (1988) has its origins in a similar concern for the ethical dilemmas posed by scientific and technological developments. As a further reminder that practical action is intimately associated with moral judgements, Hans Jonas's quest for 'an ethics for the technological age' is driven by the recognition that 'the lengthened reach of our deeds moves *responsibility*, with no less than man's fate for its object, into the center of the ethical stage' (Jonas, 1983, p. x).

There have been excursions also into the political philosophy of science, notably Joseph Rouse's *Knowledge and Power* (1987). Rouse contends that the two separate streams of philosophical reflection upon science in the 20th century, one associated with Anglo-American philosophers and the other with Continental European philosophers, although different in their approaches, nevertheless have been at one in the primacy accorded to the intellectual and epistemic characteristics of science. What have been neglected are questions such as 'why the . . . new scientific insights in the past two centuries have been so readily and extensively applicable' and 'what their social and political impact has been' (p. viii). Drawing upon 'new empiricist' philosophers of science such as Ian Hacking, Mary Hesse and Nancy Cartwright, as well as upon the ideas of other thinkers such as Michel Foucault, Martin Heidegger and Jürgen Habermas, Rouse explores why certain representations of the world assist us in manipulating and controlling it effectively. As he points out, 'it cannot be taken for granted that increased knowledge of the natural world would inevitably be usable in a significant way or that it would in fact be used' (p. viii). Much of his subsequent argument about the relationship between knowledge and power concerns science in the microworld of the laboratory, but he does address issues such as the influence of science upon technology and asks whether this is 'at least as much the transformation of scientific processes, techniques, and practices to satisfy extrascientific concerns as it is the application of scientific theories' (p. 24). Whilst his explanation of 'science as power outside the laboratory' is insightful, not least in relation to conditions for the effective technological extension of scientific knowledge, its implications for the nature of practical knowledge remain oblique rather than direct (pp. 226–36).

Sociologists, no less than philosophers, have demonstrated an interest in the nature of knowledge. Its production, validation, organization, distribution and use have all been subjects of investigation and theorizing. The relationship between knowledge systems and practical action has not, however, been a major theme of many works, one exception being the book by Holzner and Marx entitled *Knowledge Application. The Knowledge System in Society* (1979) from which a quotation at the opening of this paper was drawn. [See Article 1.2, Reader 1.] While they demonstrate convincingly that the act of applying knowledge is anything but 'a straightforward matter of calculation and rationalization, itself routine' (p. 261), and include among their examples some from the fields of science and technology, their concern is a general one and not specific to any particular domain of knowledge.

It is when we turn to psychology, however, that a developed vein of research is to be found which appears comparable to, and to some extent convergent with, work in the history and philosophy of technology and in the public understanding of science, already reviewed. This goes under various names such as 'everyday cognition' (Rogoff and Lave, 1984), 'cognition in practice' (Lave, 1988) and 'everyday understanding' (Semin and Gergen, 1990). Its specific emphases may vary, but the important role of context in cognitive activities has been a distinguishing feature of much of this work.

Rogoff and Lave's (1984) edited collection of 11 conference papers represents one landmark in this field. It brought together developmental psychologists, anthropologists, sociologists and computer scientists, all of whom were exploring the development of thinking in practical situations which included a milk-processing plant, a supermarket and a novices' ski slope, as well as more familiar educational contexts. Some of this work had its origins in cross-cultural observations that people who had difficulty with a task involving a particular skill in a formal laboratory setting could nevertheless display that skill in their everyday activities. This was not to be interpreted as evidence of a *real* cognitive capability which could be uncovered if only the context was structured aright or if, in some way, we could control for context. The laboratory is not context-free and as Rogoff and Lave point out, 'context is an integral aspect of cognitive events, not a nuisance variable' (p. 3). Their position is that 'thinking is intricately interwoven with the context of the problem to be solved' (p. 2). Following Vygotsky, they regard cognitive activity as 'socially defined, interpreted and supported' because 'central to the everyday contexts in which . . . (it) occurs is interaction with other people and use of socially provided tools and schema for solving problems' (p. 4).

The evidence assembled by the various contributors to Rogoff and Lave's volume supports the conclusion that 'thinking is a practical activity which is adjusted to meet the demands of the situation'. In everyday situations where practical action is required, 'people devise satisfactory opportunistic solutions', which, far from being 'illogical and sloppy', are 'sensible and effective in handling the practical problem' (p. 7). Their findings here have much in common with those from recent work on public understanding of science.

Drawing particularly upon the results of her Adult Math Project, but on other research also, Jean Lave's subsequent book, *Cognition in Practice* (1988), carries our understanding further. '"Cognition" observed in everyday practice', she argues, 'is distributed – stretched over, not divided among – mind, body, activity and culturally organized settings (which include other actors)'. There is no disjunction between 'theory' and 'practice'. From the study of jpfs (just plain folks, a term used with intentional irony, not least towards the condescension that 'experts' can display to 'lay' thinkers) and how, for example, they compute when shopping for groceries or when involved as dieting cooks in culinary measurements, evidence is provided of problems being defined by answers at the same time as answers are being constructed during the shaping of problems. Both problem and answer assume their form *in action* in a particular culturally structured naturalistic setting. The syncretic knowledge that is constructed in everyday practical action has similarities to technological knowledge as detailed earlier in this paper. As Rogoff has expressed it, 'the purpose of cognition is not to produce thoughts but to guide intelligent interpersonal and practical action' (1990, p. 9).

A recent extension of this vein of research is evident in studies of teachers' thinking in the specific contexts of classrooms and laboratories. According to Calderhead (1988, p. 54), we know relatively little about teachers' practical knowledge, 'the knowledge that is directly related to action', other than that it is 'qualitatively different from academic, subject matter or formal theoretical knowledge'. In a paper they entitle 'An appropriate conception of teaching science: a view from studies of science learning', Peter and Mariana Hewson (1988, p. 608) quote Lee Shulman's opinion that we need to explore in greater detail the relationship between teachers' content knowledge and the general pedagogical knowledge which they employ 'when they represent and formulate content in order to make it comprehensible to others'. Pedagogical knowledge might include 'teachers' understanding of the conceptions . . . that students of different ages and backgrounds bring to the classroom, and the strategies . . . most likely to be fruitful in reorganizing the understanding of learners'. Pamela Grossman, Susanne Wilson and Lee Shulman (1989, p. 32) in a paper based on research including that undertaken in the Knowledge Growth in a Profession Project conclude that 'one of the first challenges facing beginning teachers concerns the transformation of their disciplinary knowledge into a form of knowledge that is appropriate for students and specific to the task of teaching. The ability to transform subject matter knowledge requires more than knowledge of the substance and syntax of one's discipline; it requires knowledge of learners and learning, of curriculum and content, of aims and objectives, of pedagogy. It also requires a subject-specific knowledge of pedagogy. By drawing upon a number of different types of knowledge and skill . . . teachers translate their knowledge of subject matter into instructional representations'. This act of translation can be complex (Wilson, Shulman and Richert, 1987) and is often underestimated.

[. . .]

☐   Implications for pedagogy

The involvement of students in activities which encourage the progressive development of practical capability has pedagogical implications for all teachers, and not only for those of science. At one level, it represents a major shift in the goals of schooling which hitherto have regarded practical action, for the most part, as 'off limits'. It is, therefore, a challenge to what Barbara Rogoff (1990, p. 191) has described as 'the Euroamerican institution of schooling, which promotes an individually centred analytic approach to . . . tools of thought and stresses reasoning and learning with information considered on its own ground, extracted from practical use'. Put differently, it confronts the historic role of schools as institutions which decontextualize knowledge by requiring them to encourage its re-engagement with the practical and everyday once more.

A detailed examination of the pedagogical implications of this shift lies outside the possible scope of this paper. Donald Schön (1987) has devoted a book to explication of his concept of 'the reflective practicum', the pedagogical means by which he sees it possible to bridge the two worlds of disciplined knowledge and practical capability and this largely within specific fields of higher education. With younger students, and more generally throughout history, the traditional mode by which practical capability has been acquired has been that of apprenticeship.

It is interesting, therefore, to find psychologists who investigate 'thinking as it serves effective action in the interpersonal and physical world' adopting the metaphor of apprenticeship in their accounts of cognitive development. In so doing, they draw on Vygotsky's model for the mechanism through which social interaction facilitates cognitive development. In this, 'a novice works closely with an expert in joint problem solving in the zone of proximal development. . . . Development builds on the internalization by the novice of the shared cognitive processes, appropriating what was carried out in collaboration to extend existing knowledge and skills' (Rogoff, 1990, p. 141). On this view, 'children's cognitive development is an apprenticeship – it occurs through guided participation in social activity with companions who support and stretch children's understanding of and skills in using the tools of culture' (ibid., p. vii). And as V. John-Steiner (1985, p. 200) has noted, 'It is only through close collaboration that the novice is likely to learn what the mentor may not even know: how he or she formulates a question or starts a new project'.

One immediate reaction to this concerns the feasibility of such an apprenticeship pedagogy in a system of mass education, with classes of 30 children and often more. At a different level, whilst it is true that much practical activity, as in school technology, benefits from collaborative working, it is not always the case that the presence of a partner is helpful. Rogoff (1990, p. 163) points out that 'in some situations, the presence of a partner may serve as a distraction, requiring attention to be focused on the division of labour and on social issues rather than providing support. Some tasks may be too difficult to co-ordinate with another person, and this may be especially true for young children.' The difficulties are compounded because we lack effective procedures for analysing practical tasks in terms of their cognitive and other demands on learners. In a review of research on instructional task analysis, Michael Gardner identified performance components, knowledge structures and metacognitive knowledge as three features of tasks, but acknowledged the limited success so far achieved in the specification of general performance components of complex tasks, such as would be encountered in school technology (Gardner, 1985, p. 188). Similarly, the strands of progression in relation to the development of design and technology capability remain largely unexplored, although some beginnings have been made (Black, 1990). Of particular importance here is a conative dimension, which Mary Budd Rowe (1983), in one of the few discussions of this matter in relation to science education, calls 'fate control'. This is concerned with the sense that both social and physical events in the world are influenced by circumstances that can be uncovered, analysed and influenced, and that purposeful interventions are worthwhile and within the power of individuals or groups. If, as seems likely, a growing confidence in the ability to 'construct' problems, model possible solutions in the mind and achieve their material representation is essential to the development of practical capability, some form of apprenticeship pedagogy, offering opportunities for emulation and guided participation, however difficult to implement, would seem necessary. Because the exigencies of classroom life, including resource constraints and accountability mechanisms, are powerful determinants of teachers' everyday practices (Denscombe, 1980), the implications for institutional structures need to be considered.

☐ Implications for institutional structures

The notion of interaction is present strongly in each of the three fields of research reviewed above. Interaction between science and technology is central to any understanding of the nature of technological knowledge; interaction between science producers and users is a crucial element in the explanation of how the public 'understands' science; and interaction between learners and 'social others', both persons and contexts, is a cardinal feature of accounts of cognition in practice. Yet educational institutions too often reflect hierarchical relationships, their structure carrying, at least implicit, messages about the superiority of scientific to other understandings, with top–down and compartmentalized organizations of knowledge (Gergen and Semin, 1991, p. 2). A question arises, therefore, about the nature of institutional structures that would more effectively support the interactive relationships which seem necessary for science education to relate to praxis, and for the wider development by learners of practical capability.

Modifications to existing educational settings, involving teacher collaboration, shared resources, blocked timetables, adapted accommodation to facilitate extended group projects by students, and context-driven, rather than discipline-driven, learning have been described elsewhere (McCormick, 1987 and 1990; Smith, 1989; Medway, 1989; Layton, Medway and Yeomans, 1989; Murray, 1990). They are not recapitulated here. Similarly, a wide range of local alliances between schools and other agencies, including industrial corporations, universities and museums, with the aim of reinvigorating science education, if not relating it to practical capability, has been investigated by Myron and Ann Atkin (1989).

[. . .]

☐   **Implications for research**

Recognition of the unexplored nature of scientific knowledge application is not new. Writing over 40 years ago, R.S. Silver, as industrial scientist, wrestled with the problem of how it was that science had come to be of use in industrial enterprises. In part his concern was educational, because he was critical of the lack of explicit attempts to teach students how to go about the task 'of synthesizing abstract science into variegated patterns of experience' i.e. into specific technological applications. Students entering industry had little idea how to use their abstract knowledge in the problems presented to them. Eventually, often after a substantial time lag, many of them 'tumbled' to the nature of the process of application, 'thanks to the natural ability of human beings'. Silver's plea was for some formal study of the process of technological application in their earlier education (Silver, 1949).

Of course, knowledge of the nature of the process, even if this were available, is not the same as understanding how to implement it to achieve a desired outcome and certainly is no guarantee of an intent to act. As Hines and Hungerford (1984, p. 127) have reported in the field of environmental education, 'knowledge alone, while significantly correlated with responsible environmental action, is not sufficient to predispose individuals to attempt to remediate environmental problems'. Practical capability has crucial conative, no less than cognitive, components.

Adopting a cognitive perspective, however, the primary challenge, as Staudenmaier made clear (1985, p. 111) in relation to technological knowledge, is the nature of the process by which a design concept, interpreted broadly, becomes integrated with the specific constraints of a context to yield a particular outcome, whether an artefact of some kind or a practical action. For science education, the role of scientific knowledge and technique within this process is a matter of special interest. Although much research has been conducted on problem solving involving science (Garrett, 1986), little of this has been in relation to technological and other problems involving practical action. We lack understanding of the process and its developmental characteristics.

Recent work by historians of technology in collaboration with cognitive psychologists may be helpful here, at least in suggesting a conceptual framework for the investigation of ways in which students at various stages of their education undertake design and technology tasks. In their interpretation of invention as a cognitive process, Bernard Carlson and Michael Gorman (1990) adopt an interpretive framework involving three interrelated aspects; mental models, heuristics, and mechanical representations. Following the work of Donald Norman and others, they regard mental models as dynamic, often incomplete and unstable, visualizations which 'can be run in the mind's eye' (*ibid*., p. 390). They believe that an inventor 'possesses a mental model that incorporates his or her assumptions about how a device might eventually work' (Gorman and Carlson, 1990, p. 136). On this point, Wynne's rejection of a mental models approach to public understanding of science needs to be recalled, as well as evidence from cognition in practice (Lave, 1988) which suggests a more interactive on-going constructive process. As Carlson and Gorman acknowledge, it is not possible to acquire direct evidence of inventors' mental models and these have to be inferred from sketches, artefacts and other historical sources. Their conception differs from the stable, discrete models which the Lancaster University researchers were seeking, however, by being more transient, labile and adaptive.

By heuristics they mean 'the lines of research an inventor chooses to pursue, how she delegates work to her assistants, and how she may employ sketches, notes and models' (Carlson and Gorman, 1990, p. 392), in short, how the attempt at realization of the image is undertaken. Translated into terms of what children may do, the strategies they employ in design and technology activities would fall under this heading, as well as their relations with others, the division of labour and procedures by which work is distributed in a group, as well as sources of knowledge and techniques employed. There are interesting questions here about the perceptions which students have of what they are about and how these compare with teachers' perceptions. The progression from limited sub-task specific thinking to strategic meta-level thinking is largely unexplored (Alexander and Judy, 1988).

Although Carlson and Gorman call the third aspect of their interpretive framework 'mechanical representations', a better description might be 'material representations' because the term is intended to embrace much more than mechanical devices. It refers to whatever 'components' are deployed during design and technology activity to yield a product of some kind, whether artefact, environment or system. Interestingly, it would seem that many successful inventors build up a repertoire of 'material representations' which they draw upon time and again and which serve as a hallmark of their capability. The characteristic interaction of the three aspects, mental models, heuristics and material representations, constitutes the style of an inventor, according to Carlson and Gorman.

An invention is the technological analogue of a theory in science. Treating the act of invention as a cognitive process exposes some remarkable lacunae in our understanding of what is involved. At the same time it yields insights into ways in which science education might begin to react in productive ways with technology education in schools. Each of the three aspects of invention as a cognitive process may be influenced by what is learned in science lessons. The origins and forms of 'imagings' of what is possible, how interventions in the made world are perceived, are in large measure derived from understandings of the natural world as portrayed by science. We are not sure to what extent the heuristics by which practical realizations are achieved have features in common with heuristics of scientific activity, but the role of scientific knowledge as a resource for both the practical realization of mental models and the material representations which are employed is not in doubt, although, as we have seen, acts of translation are usually needed.

What is abundantly clear also is that a substantial research agenda awaits to be addressed at the interface of science education and praxis, and of science education and technology education in particular.

■ References

Alexander, P.A. and Judy, J.E. (1988) The interaction of domain-specific and strategic knowledge in academic performance. *Review of Educational Research*, 58 (4), 375–404.

Atkin, J.M. and Atkin, A. (1989) *Improving Science Education through Local Alliances. A Report to the Carnegie Corporation of New York*. Santa Cruz, CA: Network Publications.

Black, P. (1990) Implementing technology in the National Curriculum. In *Technology in the National Curriculum. Key Issues in Implementation*. The Standing Conference on Schools' Science and Technology and DATA, London.

Bolton, N. (1987) Technology across the curriculum. In *Technology Education Project. Paper 1*. Papers submitted to the Consultation held on 15 and 16 November 1985, St. William's Foundation, 5 College Street, York, YO1 2JF.

Calderhead, J. (1988) The development of knowledge structures to teach. In J. Calderhead (ed.) *Teachers' Professional Learning*. London: Falmer Press, pp. 51–64.

Carlson, W.B. and Gorman, M.E. (1990) Understanding invention as a cognitive process. The case of Thomas Edison and early motion pictures, 1888–91. *Social Studies of Science*, 20, 387–430.

Denscombe, M. (1980) The work context of teaching. An analytical framework for the study of teachers in classrooms. *British Journal of Sociology of Education*, 1 (3), 279–92.

Gardner, M.K. (1985) Cognitive psychological approaches to instrumental task analysis. In E.W. Gordon (ed.) *Review of Research in Education*, Vol. 12. Washington: American Educational Research Association, pp. 157–195.

Garrett, R.M. (1986) Problem solving in science education. *Studies in Science Education*, 13, 70–95.

Gergen, K.J. and Semin, G.R. (1991) Everyday understanding in science and daily life. In G.R. Semin and K.J. Gergen (eds) *Everyday Understanding. Social and Scientific Implications*. London: Sage Publications, pp. 1–18.

Gorman, M.E. and Carlson, W.B. (1990) Interpreting invention as a cognitive process: the case of Alexander Graham Bell, Thomas Edison and the telephone. *Science, Technology and Human Values*, 15 (2), 131–64.

Grossman, P.L., Wilson, S.M. and Shulman, L.S. (1989) Teachers of substance: subject matter knowledge for teaching. In M.C. Reynolds (ed.) *Knowledge Base for the Beginning Teacher*. Oxford: Pergamon Press (for the American Association of Colleges of Teacher Education).

Hewson, P.W. and Hewson, M.G.A'B. (1988) An appropriate conception of science teaching. *Science Education*, 72 (5), 597–614.

Hines, J.M. and Hungerford, H.R. (1984) Environmental education. Research related to environmental action skills. In L.A. Iozzi (ed.) *Summary of Research in Environmental Education 1971–82*. Monographs in environmental education and environmental studies, volume 2. ERIC Clearing House for Science, Maths and Environmental Education, Columbus, Ohio. ED259879.

Holzner, B. and Marx, J.H. (1979) *Knowledge Application. The Knowledge System in Society*. Boston and London: Allyn and Bacon Inc.

John-Steiner, V. (1985) *Notebooks of the Mind: Explorations in Thinking*. Alburquerque: University of New Mexico Press.

Jonas, H. (1983) *The Imperative of Responsibility. In Search of an Ethics for the Technological Age*. Chicago and London: University of Chicago Press.

Jonsen, A.R. and Toulmin, S. (1988) *The Abuse of Casuistry. A History of Moral Reasoning*. Berkeley and London: University of California Press.

Lave, J. (1988) *Cognition in Practice. Mind, Mathematics and Culture in Everyday Life*. Cambridge, New York: Cambridge University Press.

Layton, D., Medway, P. and Yeomans, D. (1989) *Technology in TVEI 14–18. The Range of Practice*. The Training Agency, Moorfoot, Sheffield S1 4PQ.

McCormick, R. (1987) *Technological Education*. Milton Keynes: The Open University Press (and other course materials for ET887/897 Teaching and Learning Technology in Schools).

McCormick, R. (1990) The evolution of current practice in technology education. Paper prepared for the NATO Advanced Research Workshop: Integrating Advanced Technology into Technology Education, 8–12 October 1990, Eindhoven.

Medway, P. (1989) Issues in the theory and practice of technology education. *Studies in Science Education*, 16, 1–24.

Murray, R. (ed.) (1990) *Managing Design and Technology in the National Curriculum: A Co-ordinated Approach*. London: Heinemann.

Nyiri, J.C. and Smith, B. (eds.) (1988) *Practical Knowledge. Outline of a Theory of Traditions and Skills*. London: Croom Helm.

Pérez-Ramos, A. (1988) *Francis Bacon's Idea of Science and the Maker's Knowledge Tradition*. Oxford: Clarendon Press.

Rogoff, B. (1990) *Apprenticeship in Thinking. Cognitive Development in Social Context*. Oxford: Oxford University Press.

Rogoff, B. and Lave, J. (eds) (1984) *Everyday Cognition: Its Development in Social Context*. Cambridge, MA: Harvard University Press.

Rouse, J. (1987) *Knowledge and Power. Towards a Political Philosophy of Science*. Ithaca and London: Cornell University Press.

Rowe, M.B. (1983) Science education: a framework for decision-making. *Daedalus*, 112 (2), 123–42.

Schön, D.A. (1987) *Educating the Reflective Practitioner*. San Francisco: Jossey-Bass Publishers.

Semin, G.R. and Gergen, K.J. (eds) (1990) *Everyday Understanding. Social and Scientific Implications*. London: Sage Publications.

Silver, R.S. (1949) Commentary: philosophy of applied science. *Research. A Journal of Science and its Applications*, 2 (4), 149–53.

Singer, P. (ed.) (1985) *In Defense of Animals*. Oxford: Blackwell.

Singer, P. (ed.) (1986) *Applied Ethics*. Oxford: Oxford University Press.

Singer, P. (1990) *Embryo Experimentation*. Cambridge: Cambridge University Press.

Smith, J.S. (ed.) (1989) *DATER 88. Proceedings of the First National Conference in Design and Technology Educational Research and Curriculum Development*. Longman Group Resources Unit, Loughborough University of Technology.

Staudenmaier, J.M. (1985) *Technology's Storytellers. Reweaving the Human Fabric*. Cambridge, MA and London: Society for the History of Technology and the MIT Press.

Wilson, S., Shulman, L. and Richert, A.E. (1987) 150 different ways of knowing: representations of knowledge in teaching. In J. Calderhead (ed.) *Exploring Teachers' Thinking*. London: Cassell.

2.4

Knowledge and Action: Science as Technology?

E. Jenkins

There are several ways in which science and technology are interdependent. The most obvious and familiar examples relate to apparatus and technique. The study of gases by chemists in the 18th century owed much to the development of effective techniques for their collection and manipulation. Likewise, in the 19th century, the establishment and measurement of a variety of 'constants' or 'standards' required the design and manufacture of new and reliable instrumentation. In the modern world, computers are perhaps the most evident feature of the interdependence of science and technology, although many other facilities, from nuclear magnetic resonance to aerial photography could also be identified.

A somewhat different kind of interdependence has less to do with technique than with ideas, constructs and modes of organization and communication. Here, science and technology are considered as two intersecting cognitive worlds that overlap in a number of ways, e.g. 'organizational structures, bodies of knowledge, traditions of practice and value systems' (Kranakis, 1991). This approach, it should be noted, is different from both the 'parasitical' notion of technology as merely applied science, and from its historiographical successor, associated with the work of Edwin Layton (1971), which asserts the independence of the scientific and technological communities.

In this article, attention is focused not on the debates about the relationships between science and technology, but on what might be called the technological aspects of scientific knowledge itself. The argument to be presented is supported by two different literatures. The first derives from work that offers an insight into a still undervalued aspect of scientific creativity, namely its craft element. The second concerns studies that have been undertaken of some of the ways in which scientific knowledge is used, not by experts, but by lay citizens addressing problems having a scientific dimension and of particular concern to them.

Everyday experience of handicrafts such as knitting and woodworking suggests that, while the requisite skills can be learnt, the very best work calls upon experience and craft knowledge that is rarely explicit. The same is true of activities such as painting or sculpture and, indeed, of any work which requires the shaping of raw materials. In discussing this work, there is an acknowledgement of a personal, tacit element, occasionally romanticized, but generally understood by reference to interactions involving the hands, eyes and mind of the skilled craftworker and the material to be shaped. Only in this way is it possible for an 'outsider' to have some understanding of what a sculptor or woodturner means by saying that an artefact doesn't 'feel right' or 'seem right'. Often, no convincing reason for this judgement can be offered since it rarely derives directly from knowledge or technique. Rather it has to do with experience and with the difficulties of trying to realize ideas, themselves not always clearly formed, in the material world. In working on, and in, this material world, problem defining and problem solving are commonly integral to each other, rather than distinct, prioritized activities, and action serves not so much to solve a problem as to (re)define it (Garner, 1990).

What is perhaps surprising is that despite much writing about the nature of scientific activity, e.g. Beveridge (1950), Polyani (1958), Hamilton (1961), Freedman (1969), Ravetz (1971) and Latour and Woolgar (1979), this insight into the nature of creativity is still largely ignored outside academic accounts of the generation of scientific knowledge. This is particularly the case with school science texts, especially those committed explicitly to the so-called 'process' approach to science teaching. Consider, as an example, *Active Science*, a publication intended for use in the early secondary years of the National Curriculum in England and Wales. This advises pupils that 'what it takes to be good at science' is 'communicating and interpreting, observing, planning investigations, investigating and making' (Coles, 1989, pp. 4–5). The concern here is not that 'doing' science is reduced to Huxley's 'clear cold logic engine', with all that this implies for students' interest and motivation, or even that the craft element of science is simply ignored. It is that the imaginative and personal elements that lie at the heart of any creative activity, including science, seem to count for nothing.

Writing about science as 'craftsman's work', Ravetz has distinguished a number of aspects of the establishment of scientific knowledge that might properly be described as craft activities. He notes that in making judgements about data generated by equipment, it is necessary to 'know', in the sense of 'have a feel for', that equipment with all its particularity, and comments that a set of quantitative readings taken from experimental apparatus 'cannot be considered independently of the interpretation put on them'. A somewhat different set of craft skills is involved in the transformation of data into information, e.g. by generating a graph or other means of representing relationships between data. Yet another craft component of scientific creativity relates to the manner in which 'pitfalls' on the road to scientific understanding can be largely avoided. This is done by 'the charting of standard paths which skirt them' and by 'each investigator becoming sensitive to the clues which indicate the special sorts of pitfalls' likely to be encountered as the work is done. The first of these is achieved by the appropriate apprenticeship training, usually involving PhD and post-doctoral studies under the guidance of experienced and successful scientists. The second is a necessary, although never a sufficient, condition for success in taking scientific knowledge beyond established limits. Both have a strong craft character, the former relying principally upon established craft practices, the latter relying critically upon tacit and personal judgements about, for example, how to proceed, or about what is or is not significant, in a field that is simultaneously known and unknown. For Ravetz, it is these and other craft elements that make possible the generation of consensible, 'objective' scientific knowledge from the intensely personal and fallible endeavour of creative scientific enquiry.

To the extent that scientific creativity relies upon these essentially craft judgements to achieve particular objectives, it has something in common with technology. There is also, of course, an important difference. Technological progress is ultimately manifest in the form of artefacts or systems, generated by operation with and upon materials that are to hand, or can be extracted or made. In contrast, science 'operates' upon the natural world in a somewhat different way. The craft judgements here relate to 'intellectually constructed things and events', and to 'problem solving on artificial objects' (Ravetz, 1971, p. 109). Nonetheless, both rely critically upon the tacit and informal knowledge that stems from apprenticeship and both can properly lay claim to being 'knowledge in action'.

A different manifestation of knowledge in action arises from the engagement of adult citizens in personal, social or other problems that have a scientific dimension. Examples, necessarily many and diverse, include parents facing the challenges presented by a child born with Down's Syndrome; elderly people seeking to manage an energy budget within a modest, fixed income; local councillors taking decisions about the handling or storage of hazardous waste; workers in the nuclear industry seeking meaningful estimations of the risks to their health; and environmental groups addressing a range of issues from aircraft noise to pollution and land conservation. Studies of the use made of scientific knowledge in these contexts (e.g. Stern and Aronson, 1984; Layton, Jenkins, McGill and Davey, 1992; MacGill, 1987; Wynne, 1990) have revealed the fundamental inadequacy of the so-called 'cognitive deficit' model of scientific literacy which presupposes that such literacy can be measured against the yardstick of orthodox science as understood within the scientific community. Instead, what has emerged is an altogether more complex and subtle picture of scientific literacy as knowledge in action – one that entails the abandonment of the notion of science as an unproblematic and unified entity which can be deployed in a straightforward manner to solve or ameliorate a range of everyday problems. In order to form the basis of action, scientific knowledge often has to be reshaped, refashioned and contextualized in ways that allow its integration with other kinds of knowledge, beliefs and judgements which are often personal and markedly context-bound.

Consider, for example, the relationship between scientific ideas about energy and the management of a domestic energy policy by the elderly. A rational approach to such management is likely to involve one or more of the following elements:

(1) Considering the possibility of moving to accommodation which makes fewer energy demands.

(2) Selecting the most cost-effective means of providing and using energy for cooking and warmth, e.g. 'off-peak' fuel tariffs; 'one-cup' kettles.

(3) Optimizing the use of devices and strategies available for controlling energy consumption, e.g. thermostats; the use of showers rather than baths; thermal insulation.

(4) Monitoring energy consumption, both *in toto* and in respect of individual appliances and activities and, where possible and appropriate, taking corrective action.

(5) Using data such as weather forecasts to anticipate and respond to consequent changes in energy needs.

However, studies have shown that an energy conservation policy based on elements such as these is unlikely to meet with much success because it fails to articulate with the everyday needs and preferences of the people concerned. There are often good reasons for not moving house (e.g. to remain near one's grandchildren or close to friends), or for preferring electricity to gas (e.g. the absence of a smell). In other words, domestic energy consumption has psychological, sociological and personal, as well as scientific, dimensions. Williams has illustrated this point with the telling comment that 'People may get warmth from fires (which is what the energy theorists believe fires are for), but in practice they *use* fires for other purposes . . . a feeling of well-being, or security, or a focal point . . . so fires have a place in a person's life that is very different from the heating engineer's view of a fire' (Williams, 1985).

In addition, attempts to articulate orthodox, established and 'universal' scientific knowledge with other forms of knowledge that are particular, local and personal, and more overtly value-laden, sometimes raise questions of trust, confidence and reliability. Not surprisingly, therefore, they often prompt a direct challenge to scientific expertise, which, in some cases, leads to its rejection in favour of alternative, more inclusive, understandings. As the debates about lead in the environment have shown only too clearly, it is possible for highly qualified experts to disagree about almost any aspect of the problem they are trying to resolve, whether this be the quality, relevance, interpretation or significance of the evidence, or even whether a particular study can properly be acknowledged as 'scientific'. Collingridge and Reeve (1986) have attributed disagreements of this kind to a breakdown in the 'autonomy, disciplinarity and . . . low level of criticism' necessary for effective scientific research and analysis, 'leading to endless technical debate rather than the hoped-for consensus which can limit arguments about policy'. Wynne (1990) offers a somewhat different insight into the difficulties of articulating scientific knowledge with action.

> '. . . ordinary life, which often takes contingency and uncertainty as normal, and adaptation to uncontrolled factors as a routine necessity, is in fundamental tension with the basic culture of science, which is premised on assumptions of manipulability and control.' (*Wynne, 1990, p. 17*)

Scientific knowledge 'in action' amid the contingency and uncertainty of the everyday world illustrates both the strength and the limitations of 'laboratory science'. More significantly in the present context, it confirms that, like technology, such scientific knowledge in action accommodates values, preferences, goals and beliefs, even methodologies and outcomes, that are integral to the issue being addressed but not necessarily consonant with those of science itself.

■ Some issues

One of the extraordinary aspects of science is the remarkable confidence that can be placed in scientific knowledge generated by the creative endeavours of fallible human beings. There are, of course, always limits to that confidence, but it would be churlish indeed not to acknowledge what has been achieved. For the most part, school science education has been concerned with introducing students to some of these achievements, implying or claiming in the process that such achievements have been brought about by the application of scientific methods. Unfortunately, in making this claim, the essentially imaginative and personal component of the scientific endeavour has all too often been ignored, both in pedagogy and in the rhetoric (Bazerman, 1988) and practice of scientific writing (Medawar, 1964). The price being paid for this may be heavy, not least in discouraging girls from studying science. However, giving more prominence to the personal and craft elements of science raises formidable pedagogical challenges, particularly at school level. As with any craft activity, techniques can be taught and learnt, and much of school laboratory teaching is directed towards this end. However, acquiring technique is not the same as 'doing science', an undertaking that requires a long and demanding apprenticeship. Many science curriculum developments, not least those of the 1960s and the 'process' science initiatives of the 1980s, have ignored this issue of apprenticeship and, in so doing, have encouraged assumptions about the nature of scientific activity that are unlikely to survive scrutiny. School science education has traditionally prioritized the cognitive and technical aspects of science. It is perhaps time to reassert and exemplify the *imaginative* and craft dimensions of one of the greatest of human endeavours.

The imaginative dimension may take a number of forms. Newton's astonishing claim that motion, not rest, is the normal state of affairs in the universe is as much a 'rape of the senses' as Galileo's heliocentric universe. Both are supreme acts of the human imagination and are rightly recognized as such. But science is full of many other, 'less grand' and workaday imaginative constructs such as the ideal gas, activation energy, oxidation state, and the recessive gene. The list is almost endless. The corollary seems to be that school science education should not present scientists as hunting for facts or formulating laws but as engaged in the building of 'explanatory structures, *telling stories* which are scrupulously tested to see if they are stories about real life' (Medawar, 1982).

The craft dimensions of scientific activity are central to this testing of stories about real life and it seems important that science education offers greater insight than hitherto into this aspect of scientific creativity. Ravetz's account of science as craft work, referred to above, suggests some of the curriculum responses necessary to bring this about. For example, students need to be given opportunities to develop 'a feel for equipment' and to recognize what this means, epistemologically as well as procedurally, when using data generated by that equipment. Likewise, students should be encouraged to challenge the commonly held belief that in many laboratory exercises 'they are merely "verifying" for themselves that certain standard effects can be produced' (Ravetz, 1971) – a reference to the manner in which the 'pitfalls' intrinsic to the generation of scientific knowledge are conveniently bypassed.

Ravetz's analysis has a number of implications for technology education. As with science education, the underlying personal, and therefore judgemental, nature of technological creativity needs to be made explicit. In some ways, this may be easier to achieve since, compared with science education, technology education is a relatively undefined field that lacks a well-established pedagogical tradition. However, the difficulty should not be underestimated, since what is involved is not simply a matter of selecting, realizing and evaluating a preferred solution to a technological problem but also an explicit recognition of the dimensions and nature of the decisions made as that problem is addressed. There may, of course, be a dilemma here. Learning to solve a problem and learning about problem-solving activity are two different activities and it is by no means obvious that both can be brought about simultaneously by the same curriculum. The parallel with science education, with its dual emphasis on learning a body of scientific knowledge and developing an understanding of what it means to be a scientist, suggests that a degree of scepticism may be appropriate.

For both science and technology education, the case for explicating the creative and judgemental elements seems to rest, at least in part, on the concern that, without such explication, science and technology are themselves misunderstood and thereby excluded from the family of humane studies. At the heart of the matter, therefore, is a fundamental question about scientific and technological creativity. Writing about the inspirational character of poetic or musical invention, Medawar has commented that the 'delight and exaltation that go with it somehow communicate themselves to others', adding that 'something *travels*: we're carried away'. In contrast, scientific discovery is 'a private event, and the delight that accompanies it, or the despair of finding it illusory, does not travel' (Medawar, 1982). It is important, not least for technology education, to ask whether technological creativity belongs with poetry and music, or with science.

What seems beyond doubt is that conventional science courses offer ample opportunity to illustrate some of the relationships between science and technology, and for students to engage in technological activities. Constructing equipment for use in a 'science' project can require students' engagement with the design, planning, making and evaluating activities that inform the technology component of the national curriculum. More use could almost certainly be made, to the mutual benefit of science and technology education, of rather more routine science teaching activities. Students, sometimes asked to design an experiment, might also be asked to design *simple* solutions to problems which, because solutions exist, cease to be problems. However, this is not necessarily the case for a student meeting the problem for the first time. Consider the case of a primary school student who, seeing gas bubbles forming on the leaves of an aquatic plant kept in sunlight in water in a jam jar, raised with her teacher the question of how these bubbles might be collected. The conventional answer to this question is far from obvious and the student's inquiry offered an excellent opportunity for the class to design, plan, execute and evaluate several solutions. In much the same way, it is instructive to ask students early in their secondary science education how they might collect the hydrogen being evolved from the reaction of zinc with dilute hydrochloric acid in a test tube. A common suggestion is to place a balloon over the mouth of the test tube. Attempts to realize this solution commonly lead to difficulties. For example, the volume of hydrogen formed is often too small to generate sufficient pressure to distend the balloon significantly, and the rate of diffusion of hydrogen through the fabric of the balloon is usually quite rapid.

The emphasis on the cognitive and technical aspects of science that has marked school science has often been justified, at least in part, by the claim that the science learnt is useful in helping future citizens to 'think scientifically' and thereby contribute productively to a range of personal, social or environmental issues that may engage their attention in a democratic society. Arguably, the informed citizen involved in a debate about a safe level of aircraft noise or the level of risk associated with a nearby waste disposal site is just as engaged with knowledge in action as the technologist for whom scientific knowledge is also a resource. In both cases, the relationships between knowledge and action are complex and the knowledge necessary for action may have to be translated from one context to another, if indeed it exists at all. Failure to recognize and accommodate the complexity of these relationships will ensure that both science and technology education are burdened with educational responsibilities that they cannot hope to meet. More generally, the emerging understanding of the dependence of knowledge upon context offers a growing challenge to the fundamental principle upon which schooling itself rests, namely, that students can be taught disembedded, 'universal' cognitive skills that are immediately available for general application in a variety of contexts.

Technology education promises a particularly interesting response to this challenge, since it presupposes that heuristics are not absolute and offers an opportunity both to revalue tacit knowledge within education and to re-affirm the usefulness of such knowledge in solving problems. It also demands teaching and learning that display many of the characteristics of apprenticeship rather than of conventional school pedagogy (see Brown, Collins and Duguid, 1989, for the notion of 'cognitive apprenticeship'). Whether this promise can be fulfilled remains to be seen.

■ References

Bazerman, C. (1988) *Shaping Written Knowledge*. Wisconsin: University of Wisconsin Press.
Beveridge, W.I.B. (1950) *The Art of Scientific Investigation*. London: Heinemann.
Brown, J.S., Collins, A. and Duguid, P. (1989) Situated cognition and the culture of learning. *Educational Researcher*, (Jan/Feb), 32–42.
Coles, M. (1989) *Active Science*. London: Collins Educational.
Collingridge, D. and Reeve, C. (1986) *Science Speaks to Power: The Role of Experts in Policy Making*. London: Frances Pinter.
Freedman, P. (1969) *The Principles of Scientific Research*. London: Pergamon Press.
Garner, S.W. (1990) Drawing and designing: the case for reappraisal. *Journal of Art and Design Education*, 9 (1), 39–55.
Hamilton, M. (1961) *Lectures on the Methodology of Clinical Research*. Edinburgh and London: Livingstone.
Kranakis, E. (1991) Science and technology as intersecting socio-cognitive worlds. In draft paper for a conference *Critical Problems and Research Frontiers in History of Science and History of Technology*, Madison, Wisconsin, 30 October–3 November.
Latour, B. and Woolgar, S. (1979) *Laboratory Life: The Social Construction of Scientific Facts*. London: Sage.
Layton, D., Jenkins, E., McGill, S. and Davey, A. (1992) *Inarticulate Science? Perspectives on Public Understanding of Science and their Implications for the Learning of Science*. London: Cassell.
Layton, E. (1971) Mirror-image twins: the communities of science and technology in nineteenth-century America. *Technology and Culture*, 12 (4), 562–80.
MacGill, S.M. (1987) *The Politics of Anxiety: Sellafield's Cancer-link Controversy*. London: Pion.
Medawar, P.B. (1964) Is the scientific paper a fraud? *The Listener*, 12 September, 377–8.
Medawar, P.B. (1982) Hypothesis and imagination. In *Pluto's Republic*. Oxford: Oxford University Press, pp. 115–35.
Polyani, M. (1958) *Personal Knowledge*. London: Routledge and Kegan Paul.
Ravetz, J.R. (1971) *Scientific Knowledge and its Social Problems*. Oxford: Clarendon Press.
Stern, P.C. and Aronson, E. (eds) (1984) *Energy Use: The Human Dimension*. New York: Freeman and Co.

Williams, D. (1985) Understanding people's understanding of energy use in buildings. In B. Stapford (ed.) *Consumers, Buildings and Energy*. Birmingham, UK: Centre for Urban and Regional Studies, University of Birmingham.

Wynne, B. (1990) *Knowledges in Context*. Background paper from five projects presented at a conference on Policies and Publics for Science and Technology, Science Museum, London, April 1990.

PART 3

General Features of Learning

3.1

Education and Thinking: The Role of Knowledge

R. Glaser

[This article by Glaser is long, technical and closely argued, so its reading is not to be tackled lightly. But, as the abstract indicates, it touches on issues that underlie any genuine attempt to understand how children might learn technology.]

◼ Abstract

A significant part of current investigation in cognitive psychology concerns the ability of people to reason, understand, solve problems, and learn on the basis of these cognitive activities. The knowledge that is accumulating should have lasting effects on improving and increasing the general use of these abilities. However, at the present time, the evidence available indicates an apparently improved capability of our schools to teach knowledge of the 'basics' without encouraging thinking and mindfulness. This article is an attempt to consider the scientific background of this dilemma. My plan here is to look at the theories that have encouraged this state of affairs and some of our attempts to cope with it. I will briefly indicate how various psychological theories have influenced the teaching of thinking: early associationistic theory of learning, notions of Gestalt theory and early work on problem solving, the pioneering work in modern cognitive psychology on information-processing models of problem solving, and more recent work that considers the interaction of acquired knowledge and cognitive process. I hope to show that abilities to think and reason will be attained when these cognitive activities are taught not as subsequent add-ons to what we have learned, but rather are explicitly developed in the process of acquiring the knowledge and skills that we consider the objectives of education and training.

91

■ Background

□ Connectionism

Early in this century, uneasiness with the failure to address the thinking and reasoning potential of human beings was evident in the reaction to E.L. Thorndike's work. He faced the charge that his psychology was mechanistic and explained adequately only the most rote kinds of learning. Nonetheless, his work appealed strongly to a generation of educators eager for pedagogical theory. His system was scientific and quantitative, buttressed with enormous quantities of data, and was down to earth in terms of its direct extrapolations to the everyday problems of education (McDonald, 1964).

Thorndike, as a theorist, did not ignore higher-level processes, but he reduced them to connectionistic conceptions. His studies fostered the development of curricula that emphasized the specificity of learning and direct experience with the skills and knowledge to be learned, because he had concluded that transfer effects were minimal. His ideas on the specificity of learning supported forms of instruction that many feared failed to encourage the development of higher levels of thinking. Thorndike's contributions assisted psychology in becoming scientific, but they tended to separate it from certain larger issues.

In contrast, John Dewey's less empirical and more philosophical approach attempted to maintain the focus on mental process. His attack on the reflex arc was significant in this regard (Dewey, 1896). The central psychological events of significance in learning and performance were 'mediated experiences' and events in relation to their adaptive function. Dewey spoke of learning in terms of aims, purposes and goals, and problem solving or intelligent action. But his was not a scientific psychology.

Despite the dominance of connectionism, interest in establishing a cognitive basis for a pedagogy that fosters thinking and reasoning in school learning has been continuously expressed by educators and researchers at least since John Dewey. Let me mention a few outstanding contributions that seem fresh today.

□ Understanding and cognitive structure

In elementary arithmetic, Thorndike's focus on collecting and strengthening S–R (stimulus–response) bonds promoted drill methods that were strongly opposed by certain educational psychologists of the time – William Brownell, in particular (cf. Resnick and Ford, 1981). Brownell's studies (1928, 1935) suggested that drill made children faster and better at 'immature' and cumbersome procedures, but failed to develop the kinds of competence that could evolve from an understanding of number concepts. To Brownell, learning arithmetic meant manipulating an integrated set of principles and patterns, and that required more meaningful instruction.

In 1940, George Katona, in *Organizing and Memorizing*, also emphasized the distinction between 'senseless' and 'meaningful' learning. Katona's thesis was that the prototype of learning is not associationistic connection, as Thorndike advocated, but the development of cognitive organization. Organizational structures enable the acquisition and preservation of facts, and the command of a large amount of specific information derives from this organization. Mechanical memorization is a limiting case that is resorted to only when a lack of inherent relations in the material being learned excludes the possibility of understanding.

In 1945, Max Wertheimer, in his book on *Productive Thinking* (1945/1959), described an insightful series of studies on problem solving in mathematics and science. His discussion of solving for the area of a parallelogram, in which he analysed the structural understanding that could facilitate transfer to new problems, is widely cited even today. Thus, in the 1930s and 1940s, the polarities of drill and practice on the one hand, and the development of understanding on the other, were apparent. This dichotomy still challenges theory and practice today.

In the late 1950s and early 1960s, behaviouristic psychology and its expression in programmed instruction strongly influenced instructional theory. Modern theories that are now contributing to the teaching of reasoning and understanding were beginning to emerge. This transition can be expressed here by personal anecdote through the contrast between my own work and the ideas of Bruner. In my writing at that time, I described the design of programmed-instruction lessons, based on the principles of Skinner's operant analysis (Taber, Glaser and Schaefer, 1965). Bruner (1964) also described elements of instruction. He talked about the sequence of instruction, the form of pacing, reinforcement, and feedback as I did. However, in contrast to my description, he also talked about the structure and form of knowledge, the representation of knowledge, and the influence of representation on the economy and generative power of acquired performance. This personal experience, on a small scale, mirrored for me the changes in psychological theory that were occurring.

☐ The persistence of older influences

Still, the utilization of older theories was widespread, and their impact and limitations are manifest today. In teaching reading, attention has been devoted to the acquisition of basic skills such as sound–symbol correspondence, decoding from print to sound, and phoneme and word recognition. The contributions of code and language approaches to instruction in beginning reading skills have been increasingly understood and have contributed significantly to the design of instructional materials and procedures. However, long-term effects of well-constructed primary curricula do not necessarily show up in the later acquisition of inferential and critical thinking skills required to comprehend text with meaning and understanding (National Assessment of Educational Progress, 1981; Resnick, 1979). Studies on the outcomes of schooling show that although elementary skills are improving, higher-level processes are being acquired less well.

In mathematics, there appears to be an increase in the performances associated with basic skill and computation, but little improvement and even a reported decline in mathematical understanding and problem solving (National Assessment of Educational Progress, 1981). The evidence is reiterated in science education. In a long-range perspective of various issues in this field, Champagne and Klopfer (1977) point out that despite the continuing philosophical commitment of science educators to scientific thinking, little of current practice adequately reflects this philosophy. Although there has been much work on defining objectives of science instruction that specify problem-solving criteria, instruction that fosters problem-solving ability and tests that assess it are far from satisfactory.

■ Some curricula for reasoning, problem solving, and learning skills

To this point, I have described past aspirations and current shortcomings. Over the past 10 to 15 years, however, certain school programs and textbooks have been designed to encourage thinking, problem solving, and abilities for learning (see Chipman, Segal and Glaser, in press; Segal, Chipman and Glaser, in press; and Tuma and Reif, 1980, for a discussion of these attempts). As I view these programs, they can be categorized as follows:

(a)   process-oriented programs

(b)   programs that use generally familiar knowledge

(c)   problem-solving heuristics in well-structured domains

(d)   logical thinking in the context of the acquisition of basic skills.

☐ Process-oriented programs

The goal of the first two programs I mention is to develop habits of reasoning and skills of learning to improve performance of a general metacognitive, self-monitoring character. It has previously been assumed that good problem solvers show more conscious awareness and use of active self-monitoring procedures than is apparent in the passive performance of poor problem solvers (Bloom and Broder, 1950). One example of a program designed to counteract this problem was developed by Whimbey and Lochhead (1980), entitled *Problem Solving and Comprehension: A Short Course in Analytical Reasoning*. The program requires thinking aloud to a partner about the steps taken in solving problems, problems like those used on intelligence, aptitude, and simple achievement tests. The partner points out but does not correct errors. The program assumes that few errors are made because of lack of knowledge of vocabulary, arithmetical facts, and so on, but rather because of errors in reasoning such as: failing to observe and use all relevant facts of a problem; failing to approach the problem in a systematic, step-by-step manner; jumping to conclusions and not checking them; and failing to construct a representation of the problem. Through carefully designed problem exercises, the program elicits procedures for reasoning and problem solving that avoid these errors.

A second example is the longer term program developed by Feuerstein, Rand, Hoffman and Miller (1980), entitled *Instrumental Enrichment: An Intervention Program for Cognitive Modifiability*. The authors of this program, like those of the preceding, attribute poor performance to general cognitive deficiencies that result in unsystematic information intake, impaired planning behaviour, inability to define problem goals, impulsive acting out, trial and error behaviour, and lack of appropriate cue discrimination and generalization. The instrumental enrichment program combines a wide variety of progressively demanding exercises with a set of didactic techniques that provide systemati- cally ordered and intentionally scheduled opportunities for reasoning and problem solving. The tasks used in the program are to some extent like psychometric and psychological laboratory tasks. Sets of such tasks comprise units that encourage cognitive activities like perceptual organization, problem representation, planning, goal analysis, and problem restructuring. This program, like Whimbey and Lochhead's, is seen as a bridge between relatively content-free exercises and thinking in curriculum content domains.

☐ **Programs that use generally familiar knowledge**

The next two programs I describe differ from those just mentioned in that they teach thinking in the context of generally familiar knowledge. Covington, Crutchfield, Davies and Olton (1974) have published a program entitled *The Productive Thinking Program: A Course in Learning to Think*. Each lesson in the program is based on an illustrated story which presents a challenging problem (such as planning a redevelopment project for a city) that the students attempt to solve. The students are led through a problem-solving process and at appropriate points are required to state the problem in their own words, formulate questions, analyse information, generate new ideas, test hypotheses, and evaluate possible courses of action. These procedures are formulated as thinking guides that are presented throughout the various lessons and problem sets.

Another program developed over the past 10 or so years is The CoRT thinking program by de Bono (in press) in England (CoRT stands for Cognitive Research Trust). The specific thinking strategies taught are like the metacognitive, self-monitoring strategies that have been already mentioned. A number of features of the program make it both similar and dissimilar to the others described here. The contents of the program are topics of interest in everyday life, such as deciding on a career, how to spend one's holiday, moving to a new house, and changing to a new job. This program emphasizes skills that are not dependent on the prior acquisition of curriculum subject matter. However, unlike Whimbey and Lochhead and Feuerstein, the CoRT program keeps away from puzzles, games, and other such abstractions.

☐   **Problem-solving heuristics in well-structured domains**

Another category of programs is concerned with teaching skill in problem solving, particularly in formal, well-structured domains like mathematics, physics, and engineering. The mathematician, George Polya (1957), and Newell and Simon (1972) are the guiding spirits. Polya recommends that explicit attention be paid to heuristic processes as well as to content. He suggests a variety of helpful ideas such as looking for analogical situations; looking for solutions to partial auxiliary problems; decomposing a problem and recombining elements; checking whether the conditions presented in a problem are sufficient, redundant, or contradictory; and working backwards from a proposed solution. He also discusses more specific procedures in mathematical problem solving such as using indirect proofs and mathematical induction. A related program is a course developed by Rubenstein (1975) called *Patterns of Problem Solving*. The instructional tactic of the book is to introduce the students to a wide range of specific problem-solving techniques that can be brought to bear on problems they encounter in their various specializations.

A similar, but somewhat different, approach is found in a book by Wickelgren (1974), entitled *How to Solve Problems: Elements of a Theory of Problems and Problem Solving*. This text is explicitly designed to improve the reader's ability to solve mathematical, scientific, and engineering problems. His assumption is that general problem-solving methods can be of substantial help to students in learning more specialized methods in a subject-matter field, and in solving problems when they do not completely understand the relevant material or the particular class of problems involved. Another program in this genre is a recent book by Hayes (1981), entitled *The Complete Problem Solver*, used to teach a college course on general problem-solving skills. This program is designed to introduce skills that improve problem solving, and at the same time provide up-to-date information about the psychology of problem solving.

☐   **Logical thinking in the context of the acquisition of basic skills**

Finally, I turn to a program that aims at fostering thinking skills in the specific context of school curricula, a contrast to the previous relatively curriculum-unencumbered programs that I have described. A program by Lipman, Sharp and Oscanyan (1979, 1980), entitled *Philosophy for Children*, attempts to do this. Their contention is that the hierarchy of basic skills to complex processes (from, for example, decoding in reading to meaningful comprehension) is so ingrained in educational philosophy and in educational research that it is difficult to conceive of the interdependence of basic skills and the skills of reasoning and thinking. Although it is believed that thinking skills are complex and basic skills more rudimentary, just the reverse may be the case. A discipline that stresses formal inquiry might be considered in the very beginning of a curriculum rather than later in the educational process. Toward this end the several parts of this program employ the procedures of philosophic logic and inquiry in the context of science, ethics, social studies, and language arts. The program designers believe that thinking is de-emphasized in education that gives either knowledge acquisition or problem-solving techniques a primary status. Lipman states that the pragmatic nature of inquiry must be made apparent in the course of acquiring knowledge and skill.

■ Comment

The above descriptions sample current practices that are evident in published programs and texts used in various educational settings. With some exceptions, I find that most of these programs place emphasis on the teaching of general processes – general heuristics and rules for reasoning and problem solving – that might be acquired as transferable habits of thinking. Also, in large part, abstract tasks, puzzle-like problems, and informal life situations are used as content. An avoidance of the complexity of subject-matter information is typical. The practical reason offered is that teachers and students would find this difficult to manage and inhibiting of the thinking processes that need to be practised and acquired. The significant aspect is that little direct connection is made with thinking and problem solving in the course of learning cumulative domains of knowledge – that is, in the context of acquiring structures of knowledge and skill that comprise the subject matter of schooling.

The deep, underlying reason for this, I believe, is a matter of theory and knowledge of human thinking. The programs that I have described are based on early theories of human cognition (some of which stem from psychometric notions of inductive reasoning) and on concepts of divergent thinking in older theories of problem solving. Others derive from early information-processing theory that explored knowledge-lean problems and that concentrated on basic information-processing capabilities humans employ when they behave more or less intelligently in situations where they lack any specialized knowledge and skill. When faced with such novel situations, they resort to general methods. But in the context of acquired knowledge and specific task structures, these methods may be less powerful; they lack the focus of domain specificity because of their wide applicability and generality (Newell, 1980). Although the general heuristic processes that humans use to solve problems have been richly described by the pioneering work of Newell and Simon (1972) and others (e.g. Greeno, 1978), this research used relatively knowledge-free problems, and as such offered limited insight into learning and thinking that require domain-specific knowledge.

In contrast, more recent work on problem solving done in knowledge-rich domains shows strong interactions between structures of knowledge and cognitive processes. The results of this newer research and theory force us to consider the teaching of thinking not only in terms of general processes, but also in terms of knowledge structure–process interactions. The feasibility of a more integrated approach is now increased by studies in developmental psychology and cognitive science in which attention has turned to cognitive process in the context of the acquisition of structures of knowledge and skill.

■ The focus on knowledge

Let me consider now this focus on knowledge. Much recent work emphasizes a new dimension of difference between individuals who display more or less ability in thinking and problem solving. This dimension is the possession and utilization of an organized body of conceptual and procedural knowledge, and a major component of thinking is seen to be the possession of accessible and usable knowledge. Evidence from a variety of sources converges on this conception: data and theory in developmental psychology, studies of expert and novice problem solving, and process analyses of intelligence and aptitude test tasks.

☐ Developmental studies

The interaction of knowledge and cognitive process has been shown in the study of memory. For example, Chi (1978) studied recall with children and adults in the standard memory for digits task and in memory for chess positions comparing high-knowledge, ten-year-old children who played tournament chess and low-knowledge adults who knew little chess. In the digit-span test, children and adults exhibited the typical result – digit span being lower for children than adults. In recall for chess positions, however, the children's memory was far superior to the adults', replicating the chess studies of Chase and Simon (1973) in which high-knowledge subjects showed better memory and encoding performance than low-knowledge individuals. This superiority is attributed to the influence of knowledge in this content area rather than the exercise of memory strategies as such. The hypothesis is that changes in the knowledge base can produce sophisticated cognitive performance. This relationship is further illustrated in Chi and Koeske's (1983) study of a single child's recall of dinosaur names. Changes in the amount and structure (the relationship between dinosaurs and identifying features) of knowledge influenced the amount of recall, and general memory strategy appeared to play a minimal role.

Next, I cite two sets of developmental studies that suggest that reasoning and problem solving are greatly influenced by experience with new information. In a recent study, Susan Carey (in press) has proposed an interesting interpretation of animistic thinking in young children. Based upon her own observations of a child's concept of 'alive', she suggests that a child's confusion about the concept of 'being alive' stems, in large part, from incomplete biological knowledge. Young children, four to seven years of age, believe that such biological functions and characteristics as eating, breathing, sleeping, and having internal organs like a heart, are primarily properties of people, not necessarily of animals. The more similar an animal is seen to be to a person, the more likely children are to judge that the animal has these attributes. Their knowledge is organized around an undifferentiated people structure, so they are as likely to say that worms have bones as to say that worms eat.

By age ten, all this has changed. Fundamental biological functions such as eating and breathing are attributed to all animals, and are differentiated from properties such as having bones. Humans become one species of mammal among many, each of which has basic similarities and differences. This change reflects, according to Carey, a reorganization of knowledge brought about by school learning and world knowledge: for four- to seven-year-olds, biological properties are organized in terms of their knowledge of human activities; for ten-year-olds, such knowledge is organized in terms of biological functions. Thus, the younger child's scant knowledge of biological functions results in the inability to justify the inclusion and exclusion of humans, animals, plants, and inanimate objects under the concept 'alive'. In older children, the acquisition of domain-specific information results in structured knowledge that is reflected in the ability to think about properties of the concept 'animate' and to reason appropriately.

Carey makes the general point that what can be interpreted as abstract pervasive changes in the child's reasoning and learning abilities are repeated as knowledge is gained in various domains. These changes come about with the acquisition of specific knowledge, and these knowledge structures comprise theories that enable different kinds of thinking. Theory changes of this kind, like those in science, are made as a wider and wider array of phenomena and problem situations must be explained. The acquisition of knowledge in some domains is more broadly applicable than in others. When knowledge structures of wide application like measurement, number concepts, and arithmetic problem-solving schema become available, learning and thinking in a variety of related domains can be influenced.

The acquisition of specific content knowledge as a factor in acquiring increasingly sophisticated problem-solving ability is also apparent in Robert

Siegler's 'rule assessment' approach to developmental change (Siegler and Klahr, 1982; Siegler and Richards, 1982). In this work, problem representation based upon appropriate information of a specific domain appears to influence task performance in a way that enables changes in inference processes. In a variant of the balance scale task studied by Inhelder and Piaget, Siegler found that five-year-olds have difficulty in solving problems because they fail to encode distance information; they concentrate solely on weight. After training in encoding distance so that it is a salient cue, children use this information to solve problems that involve a more sophisticated theory about the relationship of weight and distance in balance scale problems. Siegler's investigation required a detailed task analysis of the rule and cue knowledge required for different levels of performance. With this information, it was possible to determine the theory that guided a child's performance. Knowing what knowledge a child applied to a problem enabled the experimenter to match the child's current state of knowledge to learning events that helped the child to move to a new level of reasoning. With increasing knowledge, children could exercise complex rules that applied to a larger set of problems.

The significance of these data for the development and teaching of reasoning is that thinking is greatly influenced by experience with new information. Change occurs when theories are confronted by specific challenges and contradictions to an individual's knowledge. Siegler and Richards (1982) stated the issue clearly:

> 'Developmental psychologists until recently devoted almost no attention to changes in children's knowledge of specific content. . . . Recently, however, researchers have suggested that knowledge of specific content domains is a crucial dimension of development in its own right and that changes in such knowledge may underlie other changes previously attributed to the growth of capabilities and strategies.' (p. 930)

☐ Problem solving in experts and novices

The focus on knowledge is further evidenced in recent research on expert problem solving (Chase and Simon, 1973; Chi, Glaser and Rees, 1982; Larkin, McDermott, Simon and Simon, 1980; Lesgold, Feltovich, Glaser and Wang, 1981; Voss, Greene, Post and Penner, in press). Current studies of high levels of competence support the recommendation that an important focus for understanding expert thinking and problem solving is investigation of the characteristics and influence of organized knowledge structures acquired over long periods of learning and experience. In this endeavour, work in artificial intelligence (AI) has made significant contributions; in contrast to an earlier emphasis on 'pure' problem-solving techniques to guide a search for any problem (Newell, Shaw and Simon, 1960), this field, too, has come to focus on the structure of domain-specific knowledge. This shift in AI is characterized by Minsky and Papert (1974) as a change from what they call a power-based strategy for achieving intelligent thinking to a knowledge-base emphasis. They write as follows:

> 'The *Power* strategy seeks a generalized increase in computational power. It may look toward new kinds of computers . . . or it may look toward extensions of deductive generality, or information retrieval, or search algorithms. . . . In each case the improvement sought is . . . independent of the particular data base.'

> 'The *Knowledge* strategy sees progress as coming from better ways to express, recognize, and use diverse and particular forms of knowledge. . . . It is by no means obvious that very smart people are that way directly because of the superior power of their general methods — as compared with average people. . . . A very intelligent person might be that way because of specific local features of his knowledge-organizing knowledge rather than because of global qualities of his "thinking".' (*p. 59*)

Stimulated by this trend, Chi, Lesgold, and I have undertaken investigations to construct a theory of expert problem solving and its acquisition based on empirical descriptions of expert and novice performance in complex knowledge domains. The knowledge domains we study are physics, particularly mechanics (Chi *et al.*, 1982), and radiology, particularly the interpretation of X-rays (Lesgold *et al.*, 1981). A guiding question for us in this work is: how does the organization of the knowledge base contribute to the observed thinking of experts and novices? Our assumption is that the relation between the structure of the knowledge base and problem-solving process is mediated through the quality of the representation of the problem. We define a problem representation as a cognitive structure corresponding to a problem that is constructed by a solver on the basis of domain-related knowledge and its organization. At the initial stage of problem analysis, the problem solver attempts to 'understand' the problem by constructing an initial problem representation. The quality, completeness, and coherence of this internal representation determine the efficiency and accuracy of further thinking. And these characteristics of the problem representation are determined by the knowledge available to the problem solver and the way the knowledge is organized.

Our research suggests that the knowledge of novices is organized around the literal objects explicitly given in a problem statement. Experts' knowledge, on the other hand, is organized around principles and abstractions that subsume these objects. These principles are not apparent in the problem statement but derive from knowledge of the subject matter. In addition, the knowledge of experts includes knowledge about the application of what they know. For the expert, these aspects of knowledge comprise tightly connected schema. The novice's schema, on the other hand, may contain sufficient information about a problem situation but lack knowledge of related principles and their application. Our interpretation is that the problem-solving difficulty of novices can be attributed largely to the inadequacies of their knowledge bases and not to limitations in their processing capabilities such as the inability to use problem-solving heuristics. Novices show effective heuristics; however, the limitations of their thinking derive from their inability to infer further knowledge from the literal cues in the problem statement. In contrast, these inferences are necessarily generated in the context of the knowledge structure that the experts have acquired.

These results must be considered in the light of the work that followed the theoretical contribution of Newell and Simon. Problem-solving research proceeded to model search behaviour, and to verify that humans indeed solve problems according to basic heuristic processes such as means–end analysis. Numerous puzzle-like problems were investigated, all of which indicated that humans do solve problems according to this theoretical analysis to some degree (Greeno, 1978). However, the study of problem solving with large knowledge bases has provided a glimpse of the power of human thinking to use a large knowledge system in an efficient and automatic manner – in ways that minimize search. Current studies of high levels of competence support the recommendation that a significant focus for understanding expert thinking and problem solving and its development is investigation of the characteristics and influence of organized knowledge structures that are acquired over long periods of time.

☐   Process analysis of aptitude and intelligence

I consider now the third converging area – process analysis of aptitude and intelligence. In recent years, there has been extensive theoretical and empirical investigation of information-processing approaches to the study of intelligence and aptitude (cf. Hunt, Frost and Lunneborg, 1973; Sternberg, 1977, 1981b). Reflections on this work are relevant to our concerns here. In my research with Pellegrino and our associates on models of aptitude test performance (Pellegrino and Glaser, 1982), we have found that several interrelated components of performance differentiate high- and low-scoring individuals. One component is reflected in the speed of performance and appears to involve the management of working memory processes. A second component is conceptual knowledge of item content; low-scoring individuals with less available knowledge encode at surface feature levels, rather than at levels of generalizable concepts, which limits their inferential ability. A third component is knowledge of the solution procedures required for solving a particular task form, such as analogical reasoning. Low-scoring individuals display a weak knowledge of procedural constraints that results in procedural bugs and the inability to recover higher-level problem goals when subgoals need to be pursued.

We have speculated on the implications of these results for fostering the development of aptitudes for learning. The memory management component suggests the possibility of influencing processing skills – such as by training better methods for organizing and searching memory, as suggested in a number of the existing programs reviewed earlier. The other two components, however, concerned with conceptual and procedural knowledge, suggest a different emphasis. In contrast to process training, training related to an individual's knowledge base would involve acquiring and using conceptual information and knowledge of problem-solving constraints. In our studies, high-aptitude individuals appear to be skilful reasoners because of the level of their content knowledge as well as because of their knowledge of the procedural constraints of a particular problem form, such as inductive or analogical reasoning. This suggests that improvement in the skills of learning, such as required on aptitude and intelligence tests, takes place through the exercise of conceptual and procedural knowledge in the context of specific knowledge domains. Learning and reasoning skills develop not as abstract mechanisms of heuristic search and memory processing. Rather, they develop as the content and concepts of a knowledge domain are attained in learning situations that constrain this knowledge to serve certain purposes and goals. Effective thinking is the result of 'conditionalized' knowledge – knowledge that becomes associated with the conditions and constraints of its use. As this knowledge is used and transferred to domains of related knowledge, the skills involved probably then become more generalizable so that intelligent performance is displayed in the context of novel ('non-entrenched') situations (Sternberg, 1981a).

■ Schemata and pedagogical theories

This discussion on the significance of organized knowledge can be pulled together by introducing the theoretical concept of prototypical knowledge structures or schemata. Cognitive psychologists in accounting for various phenomena in memory, comprehension, problem solving, and understanding have found it useful to appeal to the notion of schemata. Schemata theory attempts to describe how acquired knowledge is organized and represented and how such cognitive structures facilitate the use of knowledge in particular ways.

A schema is conceived of as a modifiable information structure that represents generic concepts stored in memory. Schemata represent knowledge that we experience – interrelationships between objects, situations, events, and sequences of events that normally occur. In this sense, schemata are prototypes in memory of frequently experienced situations that individuals use to interpret instances of related knowledge (Rumelhart, 1981). People typically try to integrate new information with prior knowledge, and in many situations in which they cope with new information, much is left out so that they could never understand the situation without filling it in by means of prior knowledge. Estes (National Academy of Sciences, 1981) explains this point in the following vignette: 'At the security gate, the airline passenger presented his briefcase. It contained metallic objects. His departure was delayed.' To understand this commonplace incident, an individual must have a good deal of prior knowledge of air terminals. This prior knowledge is represented in memory by a schema that specifies the relationship between the roles played by various people in the terminal, the objects typically encountered, and the actions that typically ensue. Schema theory assumes that there are schemata for recurrent situations, and that one of their major functions is to construct interpretations of situations.

A schema can be thought of as a theory or internal model that is used and tested as individuals instantiate the situations they face. As is the case for a scientific theory, a schema is compared with observations, and if it fails to account for certain aspects of these observations, it can be either accepted temporarily, rejected, modified, or replaced. Like a theory, a schema is a source of prediction and enables individuals to make assumptions about events that generally occur in a particular situation, so that the knowledge they infer goes beyond the observations that are available in any one instance. Such prototypical structures play a central role in thinking and understanding, and the reasoning that occurs takes place in the context of these specific networks of knowledge.

Knowledge of the prototypical structures that describe problem situations is often a form of tacit knowledge present in effective problem solvers and skilled learners. Such available knowledge has been made apparent in research on children's ability to solve word problems in arithmetic. Different kinds of word problems vary in the semantic relationship that exists between quantities, and children differ in their knowledge of these categories of relations (i.e. in their knowledge about increases, decreases, combinations, and comparisons involving the sets of objects in a problem). Riley, Greeno and Heller (1983) have explicitly described categories of conceptual knowledge of problem structures that influence problem solving and learning. 'Change' and 'equalizing' categories describe addition and subtraction as actions that cause increases or decreases in some quantity. 'Combine' and 'compare' problem categories involve static relationships between quantities. For example, in change problems the initial quantity (Joe's three marbles) is increased by the action of Tom's giving Joe five more marbles. Equalizing problems involve two separate quantities, one of which is changed to equal the other; the problem solver is asked to make the amount of marbles Joe has the same as the amount Tom has. In combine problems there are two distinct quantities that do not change; Joe has three marbles and Tom has five marbles and the problem solver is asked to consider them in combination: How many marbles do Joe and Tom have altogether? Typical compare problems also describe two static quantities, but the problem solver is asked to determine the difference between them: How many more marbles does Joe have than Tom?

The influence of children's knowledge of these problem structures on problem solution is evident in studies showing that different problem categories are not equally difficult even when they require the same operations for solution. This suggests that solving a word problem requires more knowledge than just knowing the operations and having some skill in applying them.

Studies also show that young children can solve some word problems even before they have received any instruction in the syntax of arithmetic; and even after studying the formal notation, they may not translate simple word problems into equations. These studies suggest that children, in large part, base solutions on their knowledge and understanding of the prototypical semantic structure in a problem situation. Riley, Greeno, and Heller present analyses of children's problem-solving skill in which the major influence appears to be the acquisition of knowledge structures that enable improved ability to represent problem information. They also propose that knowledge of problem schemata is related to acquisition of efficient counting procedures and to more sophisticated problem-solving procedures.

The strong assumption, then, is that problem solving, comprehension, and learning are based on knowledge, and that people continually try to understand and think about the new in terms of what they already know. If this is indeed the case, then it seems best to teach such skills as solving problems and correcting errors of understanding in terms of knowledge domains with which individuals are familiar. Abilities to make inferences and to generate new information can be fostered by insuring maximum contact with prior knowledge that can be restructured and further developed (Norman, Gentner and Stevens, 1976).

The notion of schemata as theories that are a basis for learning suggests several important pedagogical principles. First, one must understand an individual's current state of knowledge in a domain related to the subject matter to be learned, and within which thinking skills are to be exercised. Second, a 'pedagogical theory' can be specified by the teacher that is different from, but close to, the theory held by the learner. Then third, in the context of this pedagogical theory, students can test, evaluate, and modify their current theory so that some resolution between the two is arrived at. Thus, the stage is set for further progression of schemata changes as the students work with, debug, and generate new theories.

When schema knowledge is viewed as a set of theories, it becomes a prime target for instruction. We can view a schema as a pedagogical mental structure, one that enables learning by facilitating memory retrieval and the learner's capacity to make inferences on the basis of current knowledge. When dealing with individuals who lack adequate knowledge organization, we must provide a beginning knowledge structure. This might be accomplished either by providing overt organizational schemes or by teaching temporary models as scaffolds for new information. These temporary models, or pedagogical theories as I have called them, are regularly devised by ingenious teachers. Such structures, when they are interrogated, instantiated, or falsified, help organize new knowledge and offer a basis for problem solving that leads to the formation of more complete and expert schemata. The process of knowledge acquisition can be seen as the successive development of structures which are tested and modified or replaced in ways that facilitate learning and thinking.

Along these lines, in his studies of learning physics, di Sessa (1982) has introduced the notion of 'genetic task analysis'. As I understand it, it is different from the usual forms of task analysis in the sense that it attempts to identify components of pre-existing theories of knowledge which can be involved in the development of more sophisticated theories – like, in learning elementary physics, the transition of naïve Aristotelian theory to Newtonian interpretations. What is important to analyse is not the logical prerequisites such as identified in a Gagné hierarchy, but rather 'genetically antecedent partial understandings' (p. 63). These understandings are genetic in the sense that they are 'pedagogical theories' which can be thought about and debugged in the course of the development of further understanding.

☐ Interrogation and confrontation

The pedagogical implication that follows from this is that an effective strategy for instruction involves a kind of interrogation and confrontation. Expert teachers do this effectively, employing case method approaches, discovery methods, and various forms of Socratic inquiry dialogue. Methods of inquiry instruction have been analysed by Collins and Stevens (1982), and their findings suggest a useful approach to the design of tutorial instructional systems. A major goal of good inquiry teachers, in addition to teaching facts and concepts about a domain, is to teach a particular rule or theory for the domain. This is done, in part, by helping the learner make predictions from and debug his or her current theory. A second goal is to teach ways to derive a rule or theory for related knowledge. The student learns what questions to ask to construct a theory, how to test one, and what its properties are. From effective teachers' protocols, Collins and Stevens prepared a detailed account of recurring strategies used for selecting cases and asking questions that confront the student with counter-examples, possibilities for correct and incorrect generalization, and other ways of applying and testing their knowledge.

Such interactive inquiry methods are powerful tools for teaching thinking in the context of subject matter. Certainly, inquiry methods are tuned to the teaching of theories; they encourage conceptual understanding, involve and therefore motivate students, and can be adapted to the needs of different students. Used with inadequate skill, however, as Collins and Stevens point out, an inquiry approach can become an inquisition that leaves many students behind, in dread of having their ignorance exposed. The method requires that a teacher be continually vigilant and keep in mind the particulars of each student's thinking.

Current approaches to inquiry teaching and theory-targeted instruction are of interest. Recent research has emphasized that people's everyday understanding of physics resembles Aristotelian rather than Newtonian thinking. Even after a physics course, many people hold to a naïve, pre-Newtonian view of basic mechanics (Champagne, Klopfer and Anderson, 1980; McCloskey, Caramazza and Green, 1980). To deal with this problem, several projects have been undertaken to provide simulated microworlds on computers so students can explore the implications of their beliefs (cf. Papert, 1980). Excellent examples have been created by diSessa (1982) and by Champagne, Klopfer, Fox, and Scheuerman (1982), who have designed computer simulations of classic physics experiments that allow students to reason about implications of their own theories and then compare events in their world to the predictions of other theories.

■ General and specific thinking skills

Now that I have emphasized teaching thinking in the context of knowledge structures and the acquisition of new knowledge, I must return to the development of general intellectual capabilities somewhat like those that are the objectives of instruction in the school programs I mentioned. In what follows, I would like to refer particularly to the self-regulatory or metacognitive capabilities present in mature learners. These abilities include knowing what one knows and does not know, predicting the outcome of one's performance, planning ahead, efficiently apportioning time and cognitive resources, and monitoring and editing one's efforts to solve a problem or to learn (Brown, 1978). These skills vary widely. Although individuals can be taught knowledge of a rule, a theory, or a procedure, if transfer of learning to new situations is a criterion, then they need to know how to monitor the use of this knowledge. Self-regulatory activities thus become important candidates for instruction, and their presence can predict student ability to solve problems and learn successfully. My hypothesis is that these self-monitoring skills can become abstracted competencies when individuals use them in a variety of literacy tasks and several fields of knowledge. They are learned as generalizations of the cognitive processes employed in daily experiences with the details of attained and new knowledge. However, these general methods may be a small part of intelligent performance in specific knowledge domains where one can rapidly access learned schemata and procedures to manipulate a problem situation. General processes may be more largely involved when an individual is confronted with problems in unfamiliar domains.

The current literature poses a dilemma between instructional emphasis on general domain-independent skills or domain-specific skills (Chipman *et al.*, in press; Segal *et al.*, in press; Tuma and Reif, 1980). It is evidenced by the emphasis of most of the school programs I described on domain-free methods and by current research on problem solving in the context of specific knowledge structures. A central issue for theory and experiment in resolving this issue will focus on the transferability of acquired knowledge and skill. There are several possibilities. First, if we believe that broad domain-independent thinking and problem-solving skills are teachable in a way that makes them widely usable, then we can adopt the tactics of general methods programs. Second, if we believe that humans for the most part show limited capability in transferring such general skills, and if knowledge structure–process interactions are powerful aspects of human performance, then training in the context of specific domains is called for. Along the general–specific dimension of these two approaches, the dilemma posed is that general methods are weak because they apply to almost any situation and will not alone provide an evaluation of specific task features that enable a problem to be solved. In contrast, skills learned in specific contexts are powerful enough when they are accessed as part of a knowledge schema, but the problem of general transfer remains.

A third possibility is that both levels of thinking can be taught as subject-matter knowledge and skill are acquired. Specific declarative knowledge and associated procedural knowledge would be learned, as well as general processes involved in using one's knowledge and skill. Recent research of this kind has been reported by Brown (in press), and she suggests a combined approach. In carrying out instruction, a student's strengths and weaknesses in learning in a particular domain could be assessed. If a student has acquired much of the specific knowledge needed for subject-matter mastery, instruction aimed primarily at general self-regulatory skills might be indicated. However, if a student shows competence in general problem-solving and self-regulatory strategies, *and* is likely to employ them to guide learning in a new area, then an emphasis on knowledge and skill specific to a domain is called for. The relative emphasis on general and specific knowledge in instruction will vary as a function of both the competence of the learner and the characteristics of the domain. This tactic seems to be a reasonable one to investigate. But rather than switching between general and specific, I would also examine a fourth possibility: teaching specific knowledge domains in interactive, interrogative ways so that general self-regulatory skills are exercised in the course of acquiring domain-related knowledge.

■ Conclusion

Psychological knowledge of learning and thinking has developed cumulatively through S–R formulations, Gestalt concepts, information-processing models, and current knowledge-based conceptions. With deepening study of cognition, current research and development is increasing the likelihood that we can move to a new level of application at which a wide spectrum of thinking skills is sharpened in the course of education and training. Few other educational possibilities beckon us to apply our energies and exploratory talents as much as this one. Teaching thinking has been a long-term aspiration, and now progress has occurred that brings it into reach. The cognitive skills developed by people in a society are profoundly influenced by the ways knowledge and literacy are taught and used. We should take heed. The task is to produce a changed environment for learning – an environment in which there is a new relationship between students and their subject matter, in which knowledge and skill become objects of interrogation, inquiry, and extrapolation. As individuals acquire knowledge, they also should be empowered to think and reason.

■ References

Bloom, B.S. and Broder, L.J. (1950) *Problem-Solving Processes of College Students: An Exploratory Investigation.* Chicago: University of Chicago Press.

Brown, A.L. (1978) Knowing when, where, and how to remember: a problem of metacognition. In R. Glaser (ed.) *Advances in Instructional Psychology* Vol. 1, pp. 77–165. Hillsdale, NJ: Erlbaum.

Brown, A.L. (in press) The importance of diagnosis in cognitive skill instruction. In S.F. Chipman, J.W. Segal and R. Glaser (eds) *Thinking and Learning Skills: Current Research and Open Questions* Vol. 2. Hillsdale, NJ: Erlbaum.

Brownell, W.A. (1928) *The Development of Children's Number Ideas in the Primary Grades.* Chicago: University of Chicago Press.

Brownell, W.A. (1935) Psychological considerations in the learning and the teaching of arithmetic. *The Teaching of Arithmetic, the 10th Yearbook of the National Council of Teachers of Mathematics.* New York: Teachers College, Columbia University.

Bruner, J.S. (1964) Some theorems on instruction illustrated with reference to mathematics. In E.R. Hilgard (ed.) *Theories of Learning and Instruction: The 63rd Yearbook of the National Society for the Study of Education* Pt. 1, pp. 306–35. Chicago: National Society for the Study of Education.

Carey, S. (in press) Are children fundamentally different kinds of thinkers and learners than adults? In S.F. Chipman, J.W. Segal and R. Glaser (eds) *Thinking and Learning Skills: Current Research and Open Questions* Vol. 2. Hillsdale, NJ: Erlbaum.

Champagne, A.B. and Klopfer, L.E. (1977) A sixty-year perspective on three issues in science education: I. Whose ideas are dominant? II. Representation of women. III. Reflective thinking and problem solving. *Science Education*, 61, 431–52.

Champagne, A.B., Klopfer, L.E. and Anderson, J.H. (1980) Factors influencing the learning of classical mechanics. *American Journal of Physics*, 48, 1074–9.

Champagne, A.B., Klopfer, L.E., Fox, J. and Scheuerman, K. (1982) *Laws of Motion: Computer-Simulated Experiments in Mechanics (the A-Machine and the Inclined Plane).* New Rochelle, NY: Educational Materials and Equipment.

Chase, W.G. and Simon, H.A. (1973) The mind's eye in chess. In W.G. Chase (ed.) *Visual Information Processing*, pp. 215–81. New York: Academic Press.

Chi, M.T.H. (1978) Knowledge structures and memory development. In R. Siegler (ed.) *Children's Thinking: What Develops?* pp. 73–96. Hillsdale, NJ: Erlbaum.

Chi, M.T.H., Glaser, R. and Rees, E. (1982) Expertise in problem solving. In R. Sternberg (ed.) *Advances in the Psychology of Human Intelligence*, Vol. 1, pp. 7–75. Hillsdale, NJ: Erlbaum.

Chi, M.T.H. and Koeske, R.D. (1983) Network representation of a child's dinosaur knowledge. *Developmental Psychology*, 19, 29–39.

Chipman, S.F., Segal, J.W. and Glaser, R. (eds.) (in press) *Thinking and Learning Skills: Current Research and Open Questions.* Hillsdale, NJ: Erlbaum.

Collins, A. and Stevens, A.L. (1982) Goals and strategies of inquiry teachers. In R. Glaser (ed.) *Advances in Instructional Psychology* Vol. 2, pp. 65–119. Hillsdale, NJ: Erlbaum.

Covington, M.V., Crutchfield, R.S., Davies, L. and Olton, R.M., Jr. (1974) *The Productive Thinking Program: A Course in Learning to Think*. Columbus, OH: Charles E. Merrill.

de Bono, E. (in press) The CoRT thinking program. In J.W. Segal, S.F. Chipman and R. Glaser (eds.) *Thinking and Learning Skills: Relating Instruction to Basic Research* Vol. 1. Hillsdale, NJ: Erlbaum.

Dewey, J. (1896) The reflex arc concept in psychology. *Psychological Review*, 3, 357–370.

di Sessa, A.A. (1982) Unlearning Aristotelian physics: A study of knowledge-based learning. *Cognitive Science*, 6, 37–75.

Feuerstein, R., Rand, Y., Hoffman, M.B. and Miller, R. (1980) *Instrumental Enrichment: An Intervention Program for Cognitive Modifiability*. Baltimore, MD: University Park Press.

Greeno, J.G. (1978) Natures of problem-solving abilities. In W.K. Estes (ed.) *Handbook of Learning and Cognitive Processes* Vol. 5. Hillsdale, NJ: Erlbaum.

Hayes, J.R. (1981) *The Complete Problem Solver*. Philadelphia: Franklin Institute Press.

Hunt, E.B., Frost, N. and Lunneborg, C. (1973) Individual differences in cognition: A new approach to cognition. In G. Bower (ed.) *The Psychology of Learning and Motivation* Vol. 7, pp. 87–122. New York: Academic Press.

Katona, G. (1940) *Organizing and Memorizing: Studies in the Psychology of Learning and Teaching*. New York: Hafner.

Larkin, J.H., McDermott, J., Simon, D.P. and Simon, H.A. (1980) Models of competence in solving physics problems. *Cognitive Science*, 4, 317–345.

Lesgold, A.M., Feltovich, P.J., Glaser, R. and Wang, Y. (1981) *The Acquisition of Perceptual Diagnostic Skill in Radiology* Tech. Rep. PDS-1. Pittsburgh, PA: Learning Research and Development Center, University of Pittsburg.

Lipman, M., Sharp, A.M. and Oscanyan, F.S. (1979) *Philosophical Inquiry: Instructional Manual to Accompany Harry Stottlemeier's Discovery* 2nd edn. Upper Montclair, NJ: Institute for the Advancement of Philosophy for Children.

Lipman, M., Sharp, A.M. and Oscanyan, F.S. (1980) *Philosophy in the Classroom* 2nd edn. Philadelphia: Temple University Press.

McCloskey, M., Caramazza, A. and Green, B. (1980) Curvilinear motion in the absence of external forces: Naïve beliefs about the motion of objects. *Science*, 210, 1139–1141.

McDonald, F.J. (1964) The influence of learning theories on education (1900–1950). In E.R. Hilgard (ed.) *Theories of Learning and Instruction: The 63rd Yearbook of the National Society for the Study of Education* Pt. 1, pp. 1–26. Chicago: National Society for the Study of Education.

Minsky, M. and Papert, S. (1974) *Artificial Intelligence*. Eugene, OR: Oregon State System of Higher Education.

National Academy of Sciences (1981) *Outlook for Science and Technology: The Next Five Years*. New York: W.H. Freeman.

National Assessment of Educational Progress (1981) *Three National Assessments of Reading: Changes in Performance, 1970–80* Rep. No. 11–R-01. Denver, CO: Education Commission of the States.

Newell, A. (1980) One final word. In D.T. Tuma and F. Reif (eds) *Problem Solving and Education: Issues in Teaching and Research* pp. 175–179. Hillsdale, NJ: Erlbaum.

Newell, A., Shaw, J.C. and Simon, H.A. (1960) A variety of intelligent learning in a general problem solver. In M.C. Yovits and S. Cameron (eds) *Self-Organizing Systems: Proceedings of an Interdisciplinary Conference* pp. 153–189. New York: Pergamon Press.

Newell, A. and Simon, H.A. (1972) *Human Problem Solving*. Englewood Cliffs, NJ: Prentice-Hall.

Norman, D.A., Gentner, D.R. and Stevens, A.L. (1976) Comments on learning schemata and memory representation. In D. Klahr (ed.) *Cognition and Instruction*. Hillsdale, NJ: Erlbaum.

Papert, S. (1980) *Mindstorms: Children, Computers, and Powerful Ideas*. New York: Basic Books.

Pellegrino, J.W. and Glaser, R. (1982) Analyzing aptitudes for learning: Inductive reasoning. In R. Glaser (ed.) *Advances in Instructional Psychology* Vol. 2, pp. 269–345. Hillsdale, NJ: Erlbaum.

Polya, G. (1957) *How to Solve It: A New Aspect of Mathematical Method* 2nd edn. Princeton, NJ: Princeton University Press.

Resnick, L.B. (1979) Theories and prescriptions for early reading instruction. In L.B. Resnick and P.A. Weaver (eds) *Theory and Practice of Early Reading* Vol. 2. Hillsdale, NJ: Erlbaum.

Resnick, L.B. and Ford, W.W. (1981) *The Psychology of Mathematics for Instruction*. Hillsdale, NJ: Erlbaum.

Riley, M.S., Greeno, J.G. and Heller, J.I. (1983) Development of children's problem-solving ability in arithmetic. In H.P. Ginsburg (ed.) *The Development of Mathematical Thinking*. New York: Academic Press.

Rubenstein, M.F. (1975) *Patterns of Problem Solving*. Englewood Cliffs, NJ: Prentice-Hall.

Rumelhart, D.E. (1981) *Understanding Understanding*. La Jolla: University of California, Center for Human Information Processing.

Segal, J.W., Chipman, S.F. and Glaser, R. (eds) (in press) *Thinking and Learning Skills: Relating Instruction to Basic Research*. Hillsdale, NJ: Erlbaum.

Siegler, R.S. and Klahr, D. (1982) When do children learn? The relationship between existing knowledge and the acquisition of new knowledge. In R. Glaser (ed.) *Advances in Instructional Psychology* Vol. 2, pp. 121–211. Hillsdale, NJ: Erlbaum.

Siegler, R.S. and Richards, D.D. (1982) The development of intelligence. In R.J. Sternberg (ed.) *Handbook of Human Intelligence* pp. 897–971. Cambridge, England: Cambridge University Press.

Sternberg, R.J. (1977) *Intelligence, Information Processing, and Analogical Reasoning: The Componential Analysis of Human Abilities*. Hillsdale, NJ: Erlbaum.

Sternberg, R.J. (1981a) Intelligence and nonentrenchment. *Journal of Educational Psychology*, 73, 1–16.

Sternberg, R.J. (1981b) Testing and cognitive psychology. *American Psychologist*, 36, 1181–1189.

Taber, J., Glaser, R. and Schaefer, H.H. (1965) *Learning and Programmed Instruction*. Reading, MA: Addison-Wesley.

Tuma, D.T. and Reif, F. (eds) (1980) *Problem Solving and Education: Issues in Teaching and Research*. Hillsdale, NJ: Erlbaum.

Voss, J.F., Greene, T.R., Post, T.A. and Penner, B.C. (in press) Problem solving skill in the social sciences. In G. Bower (ed.) *The Psychology of Learning and Motivation: Advances in Research Theory*. New York: Academic Press.

Wertheimer, M. (1959) *Productive Thinking*. New York: Harper & Row (original work published 1945).

Whimbey, A. and Lochhead, J. (1980) *Problem Solving and Comprehension: A Short Course in Analytical Reasoning* 2nd edn. Philadelphia: Franklin Institute Press.

Wickelgren, W.A. (1974) *How to Solve Problems: Elements of a Theory of Problems and Problem Solving*. San Francisco: Freeman.

3.2

Students' Conceptions and the Learning of Science

R. Driver

[This article by Rosalind Driver is set in the context of children's learning in science. Children bring their scientific conceptions to bear in the work they do in technology and it picks up on the work reviewed by Glaser in Article 3.1. It is an introductory article to a complete issue of a journal, giving a more up to date view on 'schema' than that given by Glaser.]

■ Introduction

In the early 1970s research in science education began to focus on the conceptual models that lie behind students' reasoning in particular science domains. Researchers used interviews and other interpretative techniques to investigate and describe the way in which students conceptualize a range of natural phenomena, providing intriguing insights into the child's conceptual world – a world often reflecting a compelling reasonableness, 'but air weighs nothing! How can it have weight – it just floats about' and 'when you throw a ball up your force goes into it. This wears off as the ball goes up, and gravity takes over pulling the ball down.'

At the same time concern was being expressed at the lack of understanding of scientific concepts by school and university students. It was becoming recognized that meaningful learning involved the structured organization of a knowledge system in which concepts take their meaning from the theories in which they are embedded. Methods such as concept mapping and the exploration of semantic networks were developed to probe learners' knowledge structures (West and Fensham, 1974; Stewart, 1980; Novak and Gowin, 1984). These areas of work were also of interest to cognitive psychologists studying learning in formal knowledge domains.

Nearly two decades later there is an extensive literature that indicates that children come to their science classes with prior conceptions that may differ substantially from the ideas to be taught, that these conceptions influence further learning and that they may be resistant to change. [. . .]

Why is there this growth of interest in students' conceptions in science? Undoubtedly one reason is that the findings addressed the concerns of science educators and teachers directly, illuminating problems of communication and understanding that exist at the heart of the job of teaching. However, a further reason probably lies in the contribution that studies of children's conceptions have made within an emerging 'new perspective' on learning (Osborne and Wittrock, 1983; Resnick, 1983; Gilbert and Swift, 1985; West and Pines, 1985; Carey, 1986; West, 1988). Central to this perspective is the historically important view that learning comes about through the learner's active involvement in knowledge construction. Within this broadly 'constructivist' perspective learners are thought of as building mental representations of the world around them that are used to interpret new situations and to guide action in them. These mental representations or conceptual schemes in turn are revised in the light of their 'fit' with experience (von Glasersfeld, 1989). Learning is thus seen as an adaptive process, one in which the learners' conceptual schemes are progressively reconstructed so that they are in keeping with a continually wider range of experiences and ideas. It is also seen as an active process of 'sense making' over which the learner has some control.

In as far as it views learners as architects of their own learning through a process of equilibration between knowledge schemes and new experiences, this perspective reflects and builds on the Piagetian research programme. It differs from it, however, in two significant ways. Instead of focusing on the development of general logical capabilities, the new perspective emphasizes the development of domain-specific knowledge structures. In addition, whereas the emphasis in the Piagetian research programme has been on the personal construction of knowledge through an individual's interaction with the physical environment, the new perspective also acknowledges to a greater extent the social processes in knowledge construction both at the level of the individual (Edwards and Mercer, 1987; Solomon, 1987) and within the community of scientists. The writings of Vygotsky have been increasingly influential in shaping thinking about these social and cultural influences as Bruner and Haste (1987) explain:

> '(It) is Vygotsky's view that language is a symbol system which reflects sociohistorical development. Thus the set of frameworks for interpretation available to the growing individual reflects the organizing consciousness of the whole culture . . .' (p. 9)

Learning science, therefore, is seen to involve more than the individual making sense of his or her personal experiences but also being initiated into the 'ways of seeing' which have been established and found to be fruitful by the scientific community. Such 'ways of seeing' cannot be 'discovered' by the learner – and if a learner happens upon the consensual viewpoint of the scientific community he or she would be unaware of the status of the idea.

The aim of this special issue is to review the progress made in understanding the development of children's scientific knowledge from this special constructivist perspective and to assess its application to the improvement of students' science learning. Articles in the issue give a picture of the range of current theoretical, empirical and pragmatic concerns. In general the intention is to illustrate issues in the field through accounts of current ongoing enquiries including studies of the nature, status and development of students' conceptions, theoretical perspectives on the process of conceptual change, ways of promoting conceptual change in teaching situations, and metacognition and conceptual change. In this introductory article I outline some of the main trends, research questions and unresolved problems in each of these areas and suggest possible directions for further work.

■ The nature, status and development of students' conceptions

Studies that have explored students' conceptions in depth have been undertaken in a wide range of domains in science with many hundreds of papers published, although it is notable that there have been fewer studies in biological areas than in the physical sciences. There are now a number of domains that have received considerable attention (including aspects of mechanics, light, electricity, structure of matter and photosynthesis), and from which a useful picture is emerging.

The conceptions originally documented through in-depth investigations in specific domains (for example heat and temperature (Erickson, 1979), light (Guesne, 1984), mechanics (Viennot, 1979)) have been identified in a wide range of replication studies suggesting that there may be some commonality in the models that students construct to interpret events in the natural world. This claim has been supported by planned cross-country studies including students' understanding of electricity in five European countries (Shipstone et al., 1989) and cross-cultural studies of children's conceptions of the Earth in space (Mali and Howe, 1979). While these reveal considerable commonality in the type and prevalence of the conceptions that are reported, there are indications that the different cultural influences on the development of students' conceptions may also need to be taken into account (Hewson and Hamlyn, 1983).

A further important development in the field has been the growth of studies that document the progressive evolution of children's conceptions within specific domains during the school years. For example, the study by Nussbaum (1985) of the development of children's ideas about the Earth in space revealed a sequence of conceptions; young children ascribe to a flat Earth notion, this is replaced by a notion incorporating a spherical Earth but with an absolute view of 'up and down', later the directions of up and down are construed in terms of movement away from or towards the Earth. Such cross-age studies have been undertaken in a range of domains including heat and temperature (Strauss and Stavy, 1982), material substance (Holding, 1987), air (Brook and Driver, 1989), living and non-living things (Carey, 1985). Baxter (1989) reports the results of a survey of school children's conceptions about a range of simple astronomical phenomena. He identifies features in the progression of the conceptions used by children between ages nine and 16 and indicates how these findings are being used to inform teaching in this domain. Surveys such as this suggest that children may progress in their understandings by passing through a series of intermediate notions which, though they may not be correct from a scientific point of view, may however reflect progress in children's understanding. Such studies inform the longer-term sequencing of teaching topics and provide information about the range and prevalence of prior ideas that may need to be addressed within a teaching sequence.

Marton and his colleagues in Sweden have adopted an approach to the study of conceptions that they call phenomenography. One of the assumptions made within this perspective is that due to various physical and social constraints there are a limited number of ways in which human beings conceptualize phenomena. The study reported by Linder and Erickson (1989) on tertiary students' conceptions of sound was undertaken within this phenomenographic tradition. Studies of conceptions within this perspective use as data not only evidence of the ways students conceptualize phenomena but evidence from the history of ideas in the field; an approach involving a thorough analysis of the history of scientific ideas within specific domains has provided a useful basis for the study of children's ideas by others in the field (Wiser and Carey, 1983).

Differences in methodologies have led to different claims being made about the nature and status of children's conceptions. One open question is the extent to which children's conceptions are genuinely 'theory-like', that is having a coherent internal structure and being used consistently in different contexts; this is a view articulated by McCloskey (1983) and Carey (1985). While some studies provide quite strong support for this view (Vosniadou and Brewer, 1989) others are more equivocal (Engel Clough and Driver, 1986).

Although it would be unwise to overlook various social and cultural influences on children's conceptions and their progressive development during childhood, the emerging picture does hint at children's conceptions in specific domains having much in common. If this is the case, then this has implications for a substantial research programme that could inform curriculum development in science across the school years.

■ Perspectives on the process of conceptual change

Studies of students' conceptions present us with discrete snapshots in the continual construction and reconstruction of students' knowledge. Although such studies provide valuable insights that can inform curriculum planning and the possible sequencing of ideas for teaching purposes, they do not provide information on the dynamics of change, information that is necessary as a basis for designing approaches to teaching.

The perspective on the way students' conceptions change that has received most attention from science educators, draws a parallel between students' learning and the way in which theory change has been seen to take place in science itself. Students' science learning is seen to involve a process whereby a new theory progressively replaces an earlier theory. The theory change view has been argued by Carey (1985). She distinguishes between conceptual change that involves 'weak restructuring' of learners' conceptions and 'radical restructuring', a process whereby one conceptual structure is replaced by another that differs from it in a number of ways: in the meaning of the individual concepts, the relationship between them and the domains of the phenomena it explains (a distinction also reflected in the terms 'conceptual development' and 'conceptual change' (West and Pines, 1985) and 'accretion' and 'restructuring' (Rumelhart and Norman, 1981). Instructional approaches within this view differ in the extent to which students' theory change is seen to take place from 'inside out' through the autonomous actions of the learner, or from 'outside in' through externally provided support. [. . .] Nussbaum (1989) draws on this parallel between children's learning in science and the way in which science itself proceeds. He identifies distinct perspectives within contemporary philosophies of science and comments on the implications that these perspectives may have for the way in which science lessons are conducted. In particular he considers the issue of rationality, the role of the crucial experiment and the issue of evolutionary or revolutionary change.

As with theory change in science, conceptual change in learners may be the result of many complex factors. Posner *et al.* (1982), drawing in particular on the work of Toulmin on theory change in science, proposed a model that specified that for conceptual change to take place a number of conditions need to be met. Students first need to be dissatisfied with their existing conception, then for this to be replaced a new conception has to be available that can be understood by the learner and which fits with their experience and is useful in the longer term in interpreting and predicting events. [. . .] Hewson and Thornley (1989) review the studies of classroom conceptual change that have been informed by this model and they indicate how teachers can monitor and appraise the extent to which students in their classes are adopting new conceptions.

■ Conceptual change in the classroom

Classroom-based studies that are framed by this theory change view of learning have been conducted. In some studies the strategies being adopted to promote conceptual change focus on the dissatisfaction criterion by presenting students with new, possibly surprising, experiences. The approach of challenging students' prior ideas using discrepant events has been well documented (Nussbaum and Novick, 1982). However, problems with the discrepant event approach by itself are apparent. Students can avoid seeing or responding to discrepancies. Even when a discrepancy is recognized this by itself does not necessarily enable a student to replace a prior idea with a better alternative.

Other 'experience based' interventions include carefully designed activities that encourage students to differentiate compounded notions, e.g. heat and temperature (Stavy and Berkovitz, 1980), or which make more salient those physical attributes that may not be immediately apparent. For example, Séré and Weil-Barais (1989) report positive results in a study designed to promote an understanding of the conservation of matter in which a sequence of carefully designed practical experiences were used to address aspects of students' prior conceptions.

In addition to experience with physical systems, the importance of peer group discussion as a support for conceptual change has been explored and the effectiveness of giving students opportunities through discussion to make their ideas available for reflection and review has been recognized (Champagne *et al.*, 1985). In a recent study of the effect of peer discussion on students' conceptual understanding in the context of floating and sinking, Howe *et al.* (1989) report that discussion between children with differing but inadequate views facilitates understanding and that the more advanced children were helped as much as the less advanced.

Some studies focus more on overt instructional approaches designed to help learners construct new models or conceptions. It is recognized that there are topics where students are unlikely to generate the scientific conception for themselves through exposure to critical events or peer discussion and that they require more support in the process of construction of a new theory. One way of doing this is to build on knowledge elements that the learner already has. This approach of using 'bridging analogies' that enable a student to conceptualize a situation in a new way by analogy with a system they understand has been explored with success by Clement and his co-workers (see [. . .] Clement, Brown and Zietsman (1989)). Their studies in the domain of mechanics indicate that what will work as a 'bridge' cannot necessarily be anticipated and the process of bridging as well as the identification of effective analogies requires empirical study.

Rather than building on students' existing knowledge, an alternative approach is to provide support for students in constructing an alternative theoretical system and then to consider retrospectively which view, their prior conceptions or their newly constructed theory, best fits the evidence (Rowell and Dawson, 1984). This approach has been used successfully in computer-based programmes to promote conceptual change in mechanics (White and Horowitz, 1988). In such programmes, factors such as air resistance and friction can be controlled enabling students to develop stable new conceptions concerning notions which they are then able to apply in more complex, 'real world' environments.

Classroom studies designed to promote conceptual change in a specific domain often use a range of these strategies. Approaches adopted usually provide opportunities for students to make their ideas explicit and then to challenge, extend, develop or replace these using a combination of strategies. The article by Russell, Harlen and Watts (1989) describes the approaches being used in primary school classes with regular class teachers to encourage conceptual development. The study, which was undertaken in the context of teaching and learning about change of state, reports modest conceptual development in the class as a whole, and indicates how, for individual children, the process of change is complex and piecemeal.

Although the process of restructuring of conceptions does appear to occur in an unpredictable way for individuals (Scott, 1987), studies undertaken by the Children's Learning in Science Project suggest that some generalizations can be made about the paths that a class of students will tend to take in their thinking during the restructuring process. The project developed a series of teaching schemes for secondary school students in which the students were presented with experiences that they were asked to reflect upon, interpret and test their interpretation through discussion and experimentation. The same schemes of work were used by different teachers who kept a record of the ideas students introduced using diaries and diagnostic tests. A notable feature of the data was the extent to which similar conceptual pathways with their conceptual problems, apparent conflicts and helpful experiences were identified with different classes (Brook, 1987; Johnston, 1990). This suggests that well documented studies of conceptual change in classroom settings could usefully inform curriculum development and provide teachers with a helpful map of the 'conceptual ecology' in their classrooms within specific domains.

■ Alternative perspectives on learning science

Although the theory change view of learning has so far been the most dominant influence on studies of students' classroom learning in science there are other perspectives that deserve attention by science educators.

A perspective that has been described by di Sessa (1988) as a knowledge-in-pieces view portrays intuitive knowledge as a set of context-dependent schemes. Children's science ideas are thus seen as consisting 'of a rather large number of fragments rather than one or even any small number of integrated structures one might call "theories"' (p. 52) and transition to scientific understanding involves the systematic organization of these schemes. The implications for instruction of this perspective are that it is necessary to provide students with a range of experiences within a domain and to support and encourage the systematic and coherent organization of students' interpretations of those experiences.

A further perspective is provided by those who argue for 'situated cognition', a view which holds that human beings have alternative 'ways of seeing' things that are appropriate in different contexts and social situations. (It may be quite appropriate to talk at home about closing the door to keep the cold out.) Learning science from this viewpoint involves students not so much in changing their conceptions but in learning to distinguish the contexts when particular conceptions are appropriate (Solomon, 1983).

In the complex business of classroom learning in science, it is likely that all these perspectives, 'theory-change', 'knowledge-in-pieces' and 'situated cognition' have a contribution to make. The question for science educators is not so much which model to adopt but to identify, from the evidence about children's reasoning and an analysis of the structure of the science to be taught, when each may be appropriate. Millar (1989), who provides a critique of the field, presents a number of arguments concerning the relationship between research on children's learning in science and its implications for teaching. He argues that a constructivist view of learning does not logically entail a particular model of instruction and suggests that a search for a grand theory of conceptual change to inform teaching may be misguided. Instead he argues that a more painstaking approach may be needed in which each domain has its own researched curriculum development programme.

■ Conceptual change and metalearning

The question as to whether the changes that take place in children's reasoning are accounted for solely in terms of the development of domain-specific knowledge has been disputed over the years; an issue that underlies the debate concerning the validity of the Piagetian stage theory. This issue is re-emerging again with a focus on the development of general metacognitive strategies such as reflective awareness about, and deliberate control over, cognitive functioning.

In a recent book Kuhn et al. (1988) report a series of investigations into the development of students' scientific thinking skills. The general argument that is supported by their findings is that students' prior knowledge takes the form of naïve theories that undergo successive revision in the face of new experience and information, and that the nature of this revision or co-ordination process itself undergoes developmental change. They suggest that 'a major development in scientific reasoning skill is the differentiation of theory and evidence and the elevation of the process of theory/evidence interaction to the level of conscious control' (p. 9).

Following in this general line of enquiry, Carey et al. (1989) report an investigation into children's views of science. They identify through interviews a series of levels in the understandings children have of such features as the nature of hypotheses, the nature of an experiment and the relationships between scientists' ideas and other aspects of their work. The article also reports results from a small-scale intervention study designed to develop students' understanding about the nature of scientific knowledge and inquiry.

Programmes such as these, which enhance students' understanding of the nature of the scientific enterprise in the context of developing their domain-specific conceptual understanding, highlight the point that the promotion of conceptual change involves metacognitive strategies. The case for this is further developed in [. . .] White and Gunstone (1989). They argue that for effective learning to take place, including learning that involves conceptual change – or in their terms 'belief change', then an environment where learners are encouraged to reflect on their understanding and take greater responsibility for their learning needs to be provided. They report on an action research project in Australian secondary schools that aims to promote metalearning strategies in science and other subjects.

The approaches to teaching implied by the studies described here may require teachers themselves to change their views of their role as teachers and to develop a new repertoire of strategies. The process of change in science teachers' views of their role and the implications these have for teacher education and professional support is an important associated field of work.

■ Directions for future work

The documentation of students' scientific conceptions and the way these progress is a field of work that has its roots in the ethnographic tradition with its recognition of the centrality of personal meaning and of individual and cultural differences. Yet despite this orientation, there appear to be strong messages about apparent commonalities in students' conceptions that may have implications for future directions of work in this field.

☐ Studies of conceptual progression

The evidence from a number of carefully conducted studies suggests that children's ideas within specific domains tend to follow certain conceptual trajectories. Moreover, although there is variation at the individual level and there may be specific cultural influences to be considered, the general picture is that there is much in common in the conceptual trajectories for children from different backgrounds and from different countries.

If this is the case, then research effort invested in documenting these could benefit and inform science curriculum planning at an international level. Such studies could also, in the longer term, provide the kind of information on which a developmentally based assessment programme could be built.

☐ Conceptual change

Research has now documented specific prior conceptions that act as 'critical barriers' (Hawkins, 1978) to students' further learning in science. An investment of effort into developing and evaluating specific interventions that address such critical barriers could underpin the development of instructional sequences in the future. Soundly researched effective interventions deserve wider attention.

☐ Metacognition and conceptual development

Until recently the alternative conceptions movement has directed its attention to the development of domain-specific knowledge. As studies referred to earlier suggest, the involvement and possible development of more generic metacognitive strategies in the conceptual change process is one that the field needs to entertain seriously.

☐ Teacher involvement

However productive the research programmes outlined here might be, their effectiveness in terms of enhancing students' classroom learning will be limited if the implications for teaching and learning are not adopted by teachers. This has implications not only for the dissemination of the products of research through training and professional development programmes but more fundamentally it argues for the involvement of teachers in the research programmes themselves if the divide between research evidence and current pedagogical practice is to be bridged.

Acknowledgements

I am grateful to Brian Holding, Robin Millar and Philip Scott for their comments on an earlier draft of this paper.

■ References

Baxter, J. (1989) Children's understanding of familiar astronomical events. *International Journal of Science Education*, 11 (5), 502–513.

Brook, A. (1987) Designing experiences to take account of the development of children's ideas: an example from the teaching and learning of energy. In J. Novak (ed.) *The Proceedings of the Second International Seminar: Misconceptions and Educational Strategies in Science and Mathematics* Vol. II, pp. 49–64. New York: Cornell University, Ithaca.

Brook, A. and Driver, R. (1989) *Progression in Science: The Development of Pupils' Understanding of Physical Characteristics of Air across the Age Range 5–16 Years.* Children's Learning in Science Project, University of Leeds.

Bruner, J. and Haste, H. (1987) *Making Sense: The Child's Construction of the World.* London: Methuen.

Carey, S. (1985) *Conceptual Change in Childhood.* Massachusetts: MIT Press.

Carey, S. (1986) Cognitive science and science education. *American Psychologist*, 41 (10), 1123–1130.

Carey, S. *et al.* (1989) 'An experiment is when you try and see if it works': a study of grade 7 students' understanding of the construction of scientific knowledge. *International Journal of Science Education*, 11 (5), 514–529.

Champagne, A., Gunstone, R. and Klopfer, L. (1985) Effecting changes in cognitive structures among physical students. In L. West and L. Pines (eds) *Cognitive Structure and Conceptual Change.* London: Academic Press.

Clement, J., Brown, D. and Zietsman, A. (1989) Not all preconceptions are misconceptions: finding 'anchoring conceptions' for grounding instruction on students' institutions. *International Journal of Science Education*, 11 (5), 554–565.

di Sessa, A. (1988) Knowledge in pieces. In G. Forman and P.B. Pufall (eds) *Constructivism in the Computer Age*, pp. 49–70. Hillsdale, NJ: Lawrence Erlbaum Associates.

Edwards, D. and Mercer, N. (1987) *Common Knowledge.* London: Methuen.

Engel Clough, E. and Driver, R. (1986) Consistency in the use of students' conceptual frameworks across different task contexts. *Science Education*, 70 (4), 473–496.

Erickson, G.L. (1979) Children's conceptions of heat and temperature. *Science Education*, 63 (2), 221–230.

Gilbert, J. and Swift, D. (1985) Towards a Lakatosian analysis of the Piagetian and alternative conceptions research programs. *Science Education*, 69 (5), 681–696.

Guesne, E. (1984) Children's ideas about light. *New Trends in Physics Teaching* Vol. IV. Paris: UNESCO.

Hawkins, D. (1978) Critical barriers in science learning. *Outlook*, 29, 3–22.

Hewson, M. and Hamlyn, D. (1983) *The Influence of Intellectual Environment on Conceptions of Heat.* A paper presented at the annual meeting of AERA, Montreal.

Hewson, P. and Thornley, R. (1989) The conditions of conceptual change in the classroom. *International Journal of Science Education*, 11 (5), 541–553.

Holding, B. (1987) *Investigation of Schoolchildren's Understanding of the Process of Dissolving with Special Reference to the Conservation of Matter and the Development of Atomistic Ideas.* Unpublished PhD thesis, University of Leeds.

Howe, A., Rogers, C. and Tolmie, A. (1989) Physics in the primary school; peer interaction and the understanding of floating and sinking. *European Journal of Psychology of Education* (in press).

Johnston, K. (1990) *Learning and Teaching the Particulate Theory of Matter: A Report on a Teaching Scheme in Action.* Children's Learning in Science Project, University of Leeds.

Kuhn, D., Amsel, E. and O'Loughlin, M. (1988) *The Development of Scientific Thinking Skills.* London: Academic Press.

Lunder, C. and Erikson, G. (1988) A study of tertiary physics students' conceptualization. *International Journal of Science Education*, 11 (5), 491–501.

McCloskey, M. (1983) Intuitive physics. *Scientific American*, 248, 122–130.

Mali, G.B. and Howe, A. (1979) Development of earth and gravity concepts among Nepali children. *Science Education*, 63 (5), 685–691.

Millar, R. (1989) Constructive criticisms. *International Journal of Science Education*, 11 (5), 587–596.

Novar, J.D. and Gowin, D.B. (1984) *Learning How to Learn*, Cambridge: Cambridge University Press.

Nussbaum, J. (1985) The earth as a cosmic body. In R. Driver, E. Guesne and A. Tiberghien (eds) *Children's Ideas in Science*. Milton Keynes: Open University Press.

Nussbaum, J. (1989) Classroom conceptual change: philosophical perspectives. *International Journal of Science Education*, 11 (5), 530–540.

Nussbaum, J. and Novick, S. (1982) Alternative frameworks, conceptual conflict and accommodation. *Instructional Science*, 11, 183–208.

Osborne, R. and Wittrock, M.C. (1983) Learning science: a generative process. *Science Education*, 67 (4), 489–508.

Posner, G.J., Strike, K.A., Hewson, P.W. and Gertzog, W.A. (1982) Accommodation of a scientific conception: toward a theory of conceptual change. *Science Education*, 66 (2), 211–227.

Resnick, L.B. (1983) Mathematics and science learning: a new conception. *Science*, 220, 477–478.

Rowell, J.A. and Dawson, C.J. (1984) Equilibration, conflict and instruction: a new class-oriented perspective. *European Journal of Science Education*, 7, 331–344.

Rumelhart, D.E. and Norman, D.A. (1981) Analogical processes in learning. In J.R. Anderson (ed.) *Cognitive Skills and their Acquisition*, Hillsdale, NJ: Lawrence Erlbaum Associates.

Russell, T., Harlem, W. and Watt, D. (1989) Children's ideas about evaporation. *International Journal of Science Education*, 11 (5), 566–576.

Scott, P. (1987) The process of conceptual change in science: a case study of the development of a secondary pupil's ideas relating to matter. In J. Novak (ed.) *The Proceedings of the Second International Seminar: Misconceptions and Educational Strategies in Science and Mathematics* Vol. II, pp. 404–419. Ithaca, New York: Cornell University.

Séré, M.G. and Weil-Barais, A. (1989) Physics education and students' development. In P. Adey *et al.* (eds) *Adolescent Development and School Science*. Lewis: Falmer Press.

Shipstone, D.M., Rhoneck, C.V., Jung, W., Karrqvist, C., Dupin, J.J., Johsua, S. and Licht, P. (1989) A study of students' understanding of electricity in five European countries. *International Journal of Science Education*, 10 (3), 303–316.

Solomon, J. (1983) Learning about energy: how pupils think in two domains. *European Journal of Science Education*, 5 (1), 49–59.

Solomon, J. (1987) Social influences on the construction of pupils' understanding of science. *Studies in Science Education*, 14, 63–82.

Stavy, R. and Berkovitz, B. (1980) Cognitive conflict as a basis for teaching quantitative aspects of the concept of temperature. *Science Education*, 64 (5), 679–692.

Stewart, J.H. (1980) Techniques for assessing and representing information in cognitive structure. *Science Education*, 64 (2), 223–235.

Strauss, S. (ed.) with Stavy, R. (1982) *U-shaped Behavioural Growth*. New York: Academic Press.

Viennot, L. (1979) Spontaneous reasoning in elementary dynamics. *European Journal of Science Education*, 1 (2), 205–222.

von Glaserfeld, E. (1989) Cognition, construction of knowledge and teaching. *Synthese*, 80, 121–140.

Vosniadou, S. and Brewer, W. (1989) *The Concept of the Earth's Shape: A Study of Conceptual Change in Childhood*. Manuscript. Illinois: University of Illinois.

West, L. (1988) Implications of recent research for improving secondary school science learning. In P. Ramsden (ed.) *Improving Learning: New Perspectives*. London: Kogan Page.

West, L. and Pines, A. (1985) (eds) *Cognitive Structure and Conceptual Change*. London: Academic Press.

West, L.H.T. and Fensham, P.J. (1974) Prior knowledge and the learning of science. *Studies in Science Education*, 1, 61–82.

White, B. and Horowitz, P. (1988) Computer microworlds and conceptual change: a new approach to science education. In P. Ramsden (ed.) *Improving Learning: New Perspectives*, pp. 69–80. London: Kogan Page.

White, R. and Gunstone, R. (1989) Metalearning and conceptual change. *International Journal of Science Education*, 11 (5), 577–586.

Wiser, M. and Carey, S. (1983) When heat and temperature were one. In G. Gentner and A. Stevens (eds) *Mental Models*. Hillsdale, NJ: Lawrence Erlbaum Associates.

3.3

Computers in Group Settings: Doing and Learning Mathematics

L. Healy, S. Pozzi and C. Hoyles

[A necessary aspect of much technology is working in teams, where there is dependence of one person on another. Children need to develop the ability to work in this way, but team work is also a valuable learning environment. Although this article draws on research into children working together at maths on a computer, it sheds valuable light on how children may interact when given a technological problem to solve in a similar way.]

■ Background

With the relative scarcity of computers in schools, groupwork has frequently been used to maximize access to the limited numbers of machines available (e.g. Eraut and Hoyles, 1988). We believe, however, that there is considerable potential in groupwork with computers beyond mere practical convenience. Research into the relationships between interaction and learning indicates that a multitude of factors have to be taken into consideration in any attempts to understand the conditions and processes in which groupwork with computers can be fruitful. Three aspects in particular have influenced us. These have stressed the importance of restructuring the task environment (Slavin, 1983; Aronson, Bridgman and Gellner, 1978; Johnson and Johnson, 1975), the centrality of socio-cognitive conflict and negotiating joint action (Perret-Clermont, 1980; Forman and Cazden, 1985), and the construction of formal mathematical expressions with computers (Hoyles and Sutherland, 1989; Sutherland, 1989; Hoyles, Healy and Sutherland, 1991; Hoyles and Noss, 1992). Research has also focused on the different experiences which individuals bring to any group setting, and attempts to identify the influence of intra-personal and inter-personal variables on both group processes and outcomes (e.g. Webb, 1989; Barbieri and Light, 1992), with gender and ability being identified as important variables.

■ The study

In our project – Groupwork with Computers[1] – we examined a variety of group settings;[2] that is, the interaction between the pupils, the task and the software. We attempted to characterize how groups function without a teacher and the extent to which pupils together could take responsibility for task organization and the expression of mathematical ideas, both on and off the computer. We originally set out to characterize 'good' group settings, but became aware of the need to clarify the meaning of 'good', and whether this implied fruitful in terms of group outcome or the learning of individual pupils in the group. It became apparent that these two aspects needed to be distinguished, so we now describe the first as *productivity* and the second as *effectiveness*. Within an educational context, there is an inevitable tension between productivity and effectiveness – the former being the tangible product of the group interactions and the latter, though less accessible, the underlying aim of schooling. We set out to identify how far this tension was manifested within our group settings and the characteristics of those settings where productivity and effectiveness appeared to be reflexively interrelated.

We employed a multi-site case study design, working with seven classes in six schools. Eight groups of six students (aged 9–12 years) were selected, each consisting of three girls and three boys, a girl and boy from each of high, middle and low achievement levels as assessed by their teacher. This design provided a reasonable number of groups of similar composition albeit from different class and school contexts. Each group worked through three *research tasks*, two involving Logo programming and one a computer database. In each class, the teacher introduced Logo and database activities to all the pupils and decided, in conjunction with the researchers, when the pupil group was sufficiently familiar with the software to work on the research tasks. During the research sessions (each lasting about 2½ hours) each group was given one copy of the task and three computers were made available.

In order to obtain a measure of productivity and to provide a focus for the groupwork, each research task had one *group goal* designed to be challenging for pupils in the 9–12 years age-range. The research tasks involved different aspects of mathematics, namely programming, geometry and data handling, and the group goal reflected these different mathematical areas. Productivity was scored by reference to computer products and paper-and-pencil worksheets, while individual learning – effectiveness – was measured by a series of written tests completed individually a week before, immediately after and four weeks after each task. The three research tasks are shown in Figures 1 to 3.

The figures distinguish the two crucial features of our task design. First, each task consisted of a network of mathematical components – called *global*

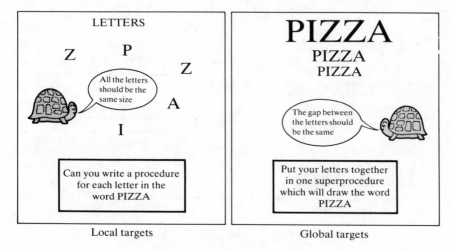

Figure 1   The letters task.

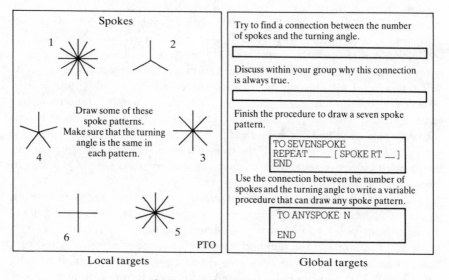

Figure 2   The Spokes task.

targets. We hypothesized that learning would be facilitated through exchange of ideas and comparison of alternative perspectives by the group as a whole around these targets. Second, each task involved a set of activities – called *local targets* – which could be shared out and constructed at the computer. The software allowed different levels of sophistication to be adopted in the process of constructing formal representations of these targets.

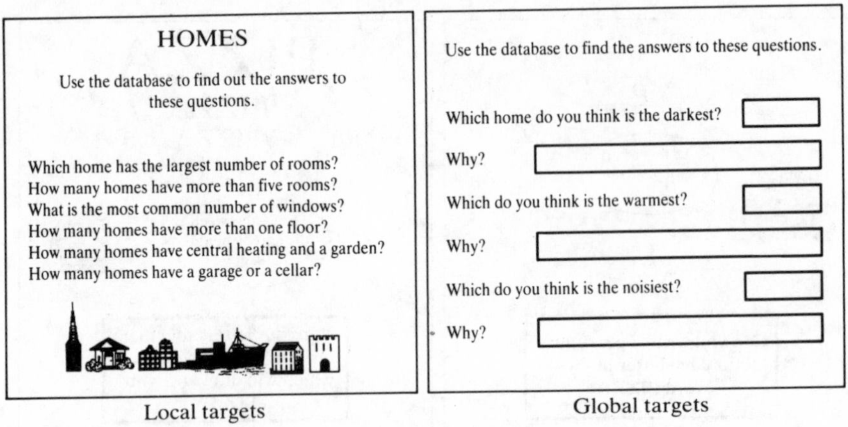

Figure 3   The homes task.

We carefully described our tasks to the pupil group after which we made no further interventions. The pupils were completely responsible for all aspects of task management: how they organized themselves, the task and the resources. Two researchers collected process data through video recordings and field notes; one recording systematically task-based interactions about the local and global targets and the other taking ethnographic notes of more general issues, including, for example, the motivation and involvement of the pupils. We interviewed all the pupils together after each task, and probed their perceptions of the task and its aims, how they thought the group had functioned and what they believed they had learnt. We also talked at length with each teacher to find out as much as possible about the group members, both individually and collectively, and the class as a whole. Finally, all the pupils in the research classes completed two questionnaires to assess perceived status in terms of 'cleverness' and popularity.

Case studies of each of the 24 group settings were constructed by weaving together the data of the group interactions whilst attempting the research task, the data from pupil and teacher interviews, the scores for group goal and individual learning and the rankings from the status questionnaire. These case studies were analysed using both qualitative and quantitative methods in order to uncover any associations of background and process factors with productivity and effectiveness.

Styles of organization

It became clear that the pupils organized themselves, the task and the available resources (worksheets and computers) in very different ways. From the case studies, we identified three styles of group organization – *collaborative*, *competitive* and *co-active*[3] – each encapsulating distinct ways in which the various aspects of the setting fitted together. In collaborative settings local targets are shared out and global targets considered by the group as a whole, working either across or away from the computers. This style most closely resembles the way the task had been planned. Where the group splits into subgroups which attempt the whole of the task separately from each other, the style of organization is termed competitive or co-active. In competitive settings, rival subgroups do *not* communicate but strive to construct the most impressive computer products. In contrast, subgroups in co-active settings, though working towards separate goals, maintain channels of communication through which help is given and task demands discussed. The three styles are illustrated schematically in Figure 4.

Collaborative styles occurred most frequently in our study (16 of 24) and invariably emerged through the actions of one or two pupils. These pupils acted as *co-ordinators*, distributing the local targets, monitoring the group progress and ensuring the group came together for the global targets. Competitive styles emerged in 25% of the settings (6 of 24) while co-active styles were rare (2 of 24). In these latter two styles, there was little or no co-ordinating activity.

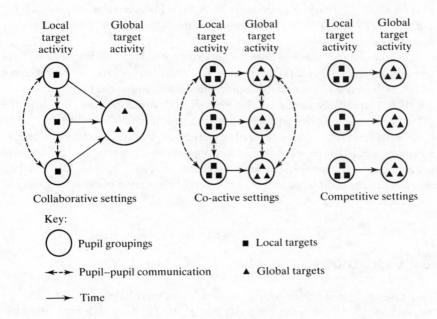

Figure 4 Styles of organization.

▓ Patterns of interaction

As well as looking at group organization, we also wanted to analyse how the groups functioned and the interrelationship of this functioning with the different styles of organization described earlier. What again emerged as important was the way individual pupils influenced the process of task solution (as well as task organization) either around the local or the global targets.

We were able to distinguish four patterns of pupil interaction, with two patterns – *directed* and *mediated* – occurring within all three organizational styles. Directed interactions are characterized by asymmetric patterns of influence with one or two pupils – *directors* – dominating both local and global targets. In contrast, in mediated interactions pupils have an equal influence over all targets with no apparent interaction roles. These two patterns represent opposite ends of a continuum in that in the former all targets are dominated and in the latter all shared. In collaborative settings only, two further patterns of interaction were identified – *navigated* and *driven*. In navigated interactions, one or two pupils – *navigators* – take control of the global mathematical issues while influence on the local targets remains evenly distributed. In contrast, in driven interactions, global target discussion is symmetric in terms of individual pupil input, but the construction of the local targets, at one computer at least, is dominated by one pupil – a *driver*. Looking at pupil roles across organizational styles and interaction patterns, we found that in collaborative settings, pupils who dominated global target interaction (the navigators or the directors) almost always acted as group co-ordinators as well. However, the majority of co-ordinators (15 of 25) took on no interaction role. These interaction patterns are summarized in Table 1.

Table 2 shows how the pupils were distributed across different patterns of interaction and organizational styles. It points to an association between collaborative style and mediated or navigated pupil interaction – an association which is particularly strong in the former case. A competitive style, on the other hand, was clearly associated with directed interactions.

Since the most common setting identified in our research was one in which a mediated pattern of interaction developed within a collaborative organization style, we present an illustration of such a setting, divided into episodes to highlight what we see as the important aspects of the functioning of the group.

▓ Case study

This case study describes a setting in which a group of Year 7 pupils (11–12-year-olds) work on the Spokes task (see Figure 2). The six pupils come from the same mathematics class, with Martin, Tony and Sophie from one tutor group, and Sharon, Graham and Kathy from another. Martin is high status, in terms of

Table 1 Interaction patterns.

Interaction pattern	Organization style	Balance of interaction	
		Local targets	Global targets
Directed	All styles	Dominated	Dominated
Mediated	All styles	Shared	Shared
Navigated	Collaborative only	Shared	Dominated
Driven	Collaborative only	Dominated	Shared

Table 2 Distribution of pupils across interaction patterns and organizational styles.

Interaction pattern	Organization style		
	Collaborative	Co-active	Competitive
Directed	6	6	24
Mediated	53	6	12
Navigated	24	n/a	n/a
Driven	13	n/a	n/a

both popularity and knowledge, and Sophie is also perceived as fairly knowledgeable, although not as popular as Martin. The other four group members are neither particularly popular nor perceived as clever.[4]

The Spokes task is based around the geometry of complete turns, and is designed to exploit pupils' understanding and use of 360° in a variety of situations. It involves exploring how to divide up a circle into different numbers of sectors, then describing a mathematical relationship between the number of sectors in a circle and the turning angle for each sector in both natural language and Logo. The individual tests associated with the task indicated that prior to the session, although all six pupils responded correctly that there were 360° in a circle, five of them could apply this idea only in a single context. One pupil – Sophie – moved beyond this context, and used 360° to calculate the angles of different sectors.

☐ **Setting up a collaborative style**

The group, co-ordinated by Martin, decides to share out the Spoke designs among three computer-based pairs:

Martin: We could share out them . . . into three
Graham: Have two each

Martin:	Yeah, have two each, have two groups on each computer, two people . . . with two things to make
All:	Yeah, OK

They have now to decide who is going to be paired with whom.

☐   Organizing into subgroups

The group discuss who will work together, eventually splitting into two single sex pairs and one mixed pair:

Sharon:	Yeah but then a girl has to go with a boy
Martin:	So! . . . That's just sexist
Graham:	I worked with Sophie last time . . . you can work with Sophie this time
Martin:	Yes I don't mind, I can work with Sophie this time
Katy:	I'm working with Sharon
Tony:	I'm going with Graham

☐   Dividing out the task

Having negotiated the pairings, Martin distributes a different activity to each pair:

Martin:	Right then who's having what? . . . One does that (*indicates Spoke designs 3 and 4*) two do that (*indicates Spoke designs 1 and 2*) and three do that (*indicates Spoke designs 5 and 6*)
Katy:	We're two
Graham:	We're one
Katy:	We do them two
Sharon:	Yeah we do them

Each pair then moves to a computer and starts to work on their assigned Spoke designs.

☐   Constructing with the computer

While interacting with the computer, the pupils in each pair develop together their own strategy to construct the rotating patterns. The first two pairs are able to work autonomously from the rest of the group. In the mixed pair who work first on a 4-Spoke design, Martin types a correct procedure straight into the editor with little need for any help from Sophie. They swap typing for the 10-Spoke, again going straight into the editor. In order to make the procedure more concise, Martin suggests the use of REPEAT and starts dictating commands to his partner. Sophie types in the commands without question until Martin suggests a turning angle provoking disagreement from Sophie:

Martin: And right 45

Sophie: 45 . . . but there aren't 10 45s in 360

Martin: What?

Sophie: There aren't 10 45s in 360

Martin: Oh yeah . . . Sharon, can we borrow your calculator?

The pair then used a calculator to find the correct angle, and their procedures are shown in Figure 5.

The boy-pair first work on the 5-Spoke with Graham typing the commands straight into the editor, using a turning angle of 85. After trying out the procedure, Tony takes over the typing, changing 85 to 75. Graham now completes and tries out the procedure. He is happy with the screen results but Tony is less sure:

Graham: Yeaahhh

Tony: Yeah . . . looks a bit wrong though . . . it's gone a bit wonky there

Graham: Who cares, we've got to do 8-Spokes now

Graham, once again typing straight into the editor, constructs a procedure for the 8-Spoke with a turning angle of 45. Their final procedures are shown in Figure 6.

```
TO 4-SPOKE
PD FD 200 BK 100 RT 90 FD 100 BK
200
END
```

```
TO 10-SPOKE
PD FD 200 BK 100
REPEAT 10 [RT 36 FD 100 BK 100]
END
```

Figure 5 Martin and Sophie's procedures.

```
TO 5-SPOKE
FD 100 BK 100 RT 70 FD 100
BK 100 RT 70 FD 100 BK 100
RT 70 FD 100 BK 100 RT 70
FD 100 BK 100
END
```

```
TO 8-SPOKE
FD 200 BK 400 FD 200 RT 90
FD 200 BK 400 FD 200 RT 45
FD 200 BK 400 FD 200 RT 90
FD 200 BK 400 FD 200
END
```

Figure 6 Tony and Graham's procedures.

```
TO 12-SPOKE
REPEAT 12 [FD 100 BK 100 RT 30]
END
```

```
TO 3- SPOKE
REPEAT 3 [FD 100 BK 100 RT 120]
END
```

Figure 7   Sharon and Kathy's procedures.

☐   **Obtaining help from the group**

Kathy and Sharon are working on a 3-Spoke design in direct drive with Kathy
typing. Sharon realizes that the design has a repetitive structure. She cannot
recall the syntax of the REPEAT command and asks for help of the group as a
whole:

> **Sharon:**   Hey everybody, how do you do a REPEAT if you want to do forward and
> back and then right?

Martin joins the subgroup briefly to give help but without affecting the girls'
general approach. The pair know the correct angle for the 3-Spoke, but when it
comes to their second design, a 12-Spoke, Sharon is less sure:

> **Sharon:**   I think we're going to have to do some sums . . . er how many 12s in 36?
> **Kathy:**   . . . 3
> **Sharon:**   3 . . . 30 degrees I think

This pair's completed procedures are shown in Figure 7.

☐   **Sharing findings from the computer work**

Having completed their two designs, each pair presents the results of their work
to the group by filling in the number of spokes and the turning angles in the
table provided (see Figure 8).

☐   **Constructing the mathematical relationship**

Martin invites the group to reassemble in order to discuss the global targets (see
Figure 2) using their completed table as data. Sharon reads out the first
question:

	NUMBER OF SPOKES	TURNING ANGLE
1 ✳	12	30
2 人	3	120
3 ✳	8	45
4 ✕	5	75
5 ✳	10	36
6 ┼	4	90

Figure 8   Pupil group's table of results.

Sharon: Try to find a connection between the number of spokes and the turning angle . . . and write your answer below there's something about it like it could be . . . the number is times by 4 add 4

The pupils start by brainstorming – generating and trying out ideas and checking each other's suggestions:

Tony: 10 is 36 now what would that be . . . that could be times 3 plus 6 . . . couldn't it . . .

Sharon: . . . Yeah could be times 3 plus 6 that's right

Sophie: That isn't

Graham: That's 30, 3 times 3 plus . . . no that doesn't work . . .

Graham notices a semi-quantitative relationship between the numbers in the table, which Sharon elaborates by referring to the Spoke designs. Graham is then ready to offer the group a general connection:

Graham: Right can I tell you something . . . the higher those are (*points at turning angles column*) the less that these are (*points at the number of spokes column*) . . .

Sharon: . . . you see that that's got to have more fitted in there, so they're going to have less degrees, that'll be loads of degrees 'cos there's only three spokes

Graham: so what it is that that's the bigger number will be the less number of turns

Martin is worried that, though correct, this idea does not provide a general method for calculating the turning angle:

> Martin: Yeah, that's all very well to say Graham but that's not what we've got to find out . . . Personally I think you have to use that, I think you have to use these numbers (*points to number of spokes column*) to get to that (*points to turning angle column*), I know that . . . right . . . so you have to get from that number to that number

The pupils take up Martin's suggestion, and all return to brainstorming. Graham has the idea of doubling, but while this idea is investigated, Sophie notices a connection with 360. She repeats herself three times — by the third time singing — but no-one takes any notice.

> Graham: And what it is . . . it wouldn't be . . .
> Sophie: (*quietly at the same time as Graham*) 4 times 90 is 360
> Graham: . . . doubling it would it . . . no
> Tony: No 'cos look 10 add 10 . . . doubling 10 gives you 20 and it's 36
> Sophie: (*while Tony is speaking*) 4 times 90 is 360
> Graham: Same as that
> Sophie: (*sings*) 4 times 90 is 360
> Tony: Double 12 is 24
> Kathy: Double that couldn't come to that

Although Sophie's idea is not taken up immediately, shortly afterwards Sharon makes the same 'discovery' and is confident that a general relationship has been uncovered:

> Martin: You have to do 3, let's do 3, you have to get somewhere from 3 to 120
> Kathy: That could get to that easily . . . I mean a 10 could get to that or 5 . . . 120
> Martin: Yeah but we don't want . . .
> Sharon: 120 times 3, 12 times 30 equals 360 . . .
> Sophie: Yeah . . . four times 90 . . .
> Sharon: . . . 3 times 120 . . . equals 360 . . . right, they all equal 360 if you times them

Martin is not yet sure of the relationship suggested by the two girls and when it is explained to him, he dismisses their ideas as too obvious:

> Martin: Yeah but times them by what?
> Sophie: These numbers
> Sharon: 8 times 45 equals 360
> Martin: Well they're going to aren't they 'cos in a full turn there are 360 degrees

☐   **Debugging an error**

Although previously rejecting the connection offered by the girls, Martin uses a variation of their idea to identify an incorrect angle (the 75° for the 5-Spoke design). In the face of Tony's opposition to his proposed change (Tony was one of the pair who had produced this result), he explains how he calculated the correct angle, but does not yet express his idea as a general rule:

Martin:	Right that's all the degrees for full turn so in the end they're all 360 . . . Right ahem . . . I can't find a single thing . . . Ahem . . . I have found a flaw in this plan that should be 72 (*points to 5-Spoke on table*)
Tony:	No it shouldn't
Martin:	Yes it should, it should be 72 . . . You'll have to go and change it, it should be 72
Tony:	Why?
Martin:	Well 'cos there is 5 spokes right and if you do 360 and then you divide that by 5 you get 72
Graham:	Well try it with this one first (*points to 12-Spoke*)
Martin:	I've tried it with all of them

☐ **Constructing a formal representation**

Graham asks Martin to explain why their turning angle is wrong. Martin tries to describe a general method and as he speaks he actually writes down an algebraic generalization ($N = 360/C$). When Graham asks what 'C' stands for, Martin suggests using the computer. The whole group assemble around one computer, and Martin is able to both communicate and validate his idea:

Graham:	Now what was your explanation, why did you say it was 72?
Martin:	Well . . . Look 'cos when you do 360 right and then you divide it by . . . divide it by . . .
Martin:	Divide it by the number of spokes . . . 'C'
Sophie:	5
Graham:	What's that?
Martin:	'C' is . . . you know . . . just say . . . 4 . . . let's just do it on the computer

☐ **Consolidating**

The group then reassembles at the table and Sharon reiterates the generalization she suggested earlier with Sophie. Now Martin takes this on board but checks it empirically by reference to pairs of numbers rather than by reference to the algebraic expression he had previously constructed. Through Martin's mediation the group finally reaches a consensus with each member contributing to the expression and explanation of the generalization:

Sharon:	. . . It could be times that number by that number it will make 360 . . . that's the connection we found that ages ago . . . so write it down . . . that's all you're going to get

Martin: . . . If you times that by that you get that, if you times that by that
                     you get 360, if you times that by that you get 360, if you times
                     that by that you get 360, if you times that by that you get 360
Tony/Graham: If you times that by that you get 360
Tony: Which is the total degrees of a full circle

□ **Writing down**

Martin co-ordinates the writing down of the group's 'agreed' answer to the first
global target question (see Figure 2) and the group goes on to consider the
remaining questions. Their answers are shown in Figure 9. For the Logo
procedure for a 7-Spoke design the pupils use a calculator to find the turning
angle and approximate this to 51° in their answer. Their misinterpretation of
the last question is of interest. We wanted them to write a general procedure
with a variable input but they saw the task as writing a procedure of a particular
'any other' spoke.

Nevertheless, with the one exception above, the group completes all

Try to find a connection between the number of spokes
and the turning angle

~~There is no formula~~ you times the number of spokes
by the Turning angle and it
should always equall 360°

Discuss within your group why this connection is always
true

It's true because it should end where
it started and therefor should end
up at ~~306~~ 360°

Finish this procedure to draw a seven spoke pattern

```
TO SEVENSPOKE
REPEAT _7_ [ SPOKE RT _51_ ]
END
```

Use the connection between the number of spokes and
the turning angle to write a variable procedure that can
draw any spoke pattern

```
TO ANYSPOKE :N   TO  ANYSPOKE
                   REPEAT 9[FD 100 BK100 RT 40]
END
```

Figure 9 Pupil group's written products.

aspects of the task correctly and obtains a high scoring group goal. Additionally, four pupils – Sharon, Martin, Tony and Graham – show progression in their understanding of 360°, applying their knowledge correctly and flexibly in both the post- and delayed-post-tests. We therefore designate this group setting as both productive in terms of outcome and effective in terms of individual learning.

■ Discussion

Our aim in this research was to characterize good group settings incorporating computers and to identify the conditions under which these emerge. In the case study described here, there was no tension between the twin goals associated with groupwork in a school context, namely group outcome and individual learning – the pupils interacted with the task and the software in ways that were both productive and effective. So why was this group setting successful? Drawing on our analysis of all the 24 case studies, we are able to suggest some explanations.

First, we looked at what the pupils brought to the setting to see if any factors outside the interactions around the task could be seen to structure the functioning of a group in a significant way. Contrary to our expectations, we found no evidence of any positive influence on the group setting of extensive prior groupwork and/or software use in the classroom – though it must be stressed that all our groups had reached what was deemed as an appropriate level of competence with the software. Prior task-related mathematical knowledge also proved to be unrelated to both productivity and effectiveness, as did the existence of any marked friendships. One condition did however need to be met for any group to function well – *the absence of overt antagonism*. Settings were unlikely to be successful in terms of either productivity or effectiveness where there was obvious hostility – particularly between girls and boys. *Age* was also important with older groups more able to manage the complexity of task, materials and human resources. In contrast, the younger pupils in our sample (9/10-year-olds) found it difficult to resolve their differences and exhibited a tendency to seek help from outside the group. Productivity but not effectiveness tended to be low in these cases, possibly due to a tension between these twin goals. In the group setting described in our case study, antagonism was not apparent, although neither were the pupils particular friends. Additionally, it is evident from the excerpts given that this group of 12-year-olds was able to function independently, make decisions and discuss outcomes.

The necessary background conditions having been satisfied, we now consider the way the case study group worked together. They shared out the task as intended, worked in subgroups on the computer and then came together as a group to discuss the more general mathematical issues and agree on their written products. This pattern of working exemplifies a collaborative style in which all interactions both on and away from the computers are mediated. Martin played an important coordinating role – indeed in all collaborative settings the *emergence of a pupil to take on and be accepted in this role of coordinator was crucial*. We looked to see if it was possible to pull out any defining characteristics of these pupil coordinators. We found that they were not necessarily popular, knowledgeable or designated by the teacher as high achievers and across all the settings, we have examples of both boys and girls taking on the role. The one attribute they all do have in common, however, was a reputation of '*cleverness*' among the whole class.

It is evident that through Martin's coordination, a *mutual interdependence* among the pupils was set up through sharing out the Spoke designs and coming together again to discuss the general relationship. Research into cooperative learning methods has indicated that such structured interdependence can have positive motivational effects leading to social and academic benefits (Slavin, 1983; Aronson, Bridgman and Gellner, 1978; Johnson and Johnson, 1975). We found that collaborative settings were likely to be productive, but not necessarily effective. We suggest that three further aspects contributed to our case study group's success in this latter area.

Firstly, during the global target activity, there was *negotiated interaction* in which pupils – with the table of their data as a focus – reflected together upon the results of the computer activity. During their discussion, the pupils communicated different ideas and considered each others' perspectives about proposed general mathematical relationships. Approaches within social cognition have emphasized these forms of interaction as central to learning (e.g. Perret-Clermont, 1980; Forman and Cazden, 1985). More specifically, within mathematics education, it is argued that negotiated interaction helps pupils distance themselves from the situated nature of mathematical experiences and encourages active pupil involvement in mathematical inquiry (see Bartolini-Bussi, 1991, and Cobb, 1990).

Secondly, during the local target activities, the pupils, within their computer-based pairs, negotiated their constructions with each other and the computer. Through this mediated pattern of interaction each pupil could develop a *sense of ownership* over mathematical strategies, while leaving room for giving and receiving help from other group members. Thirdly, the negotiated interaction occurred once pupils had been given the opportunity to *formally express* these mathematical ideas with the computer – constructions which were frequently more sophisticated and precise than those which were either oral or written. When the group came together, the pupils were able to reflect on these strategies, and where there were differing points of view, the formal language of the computer offered a way to share ideas. Thus, when Martin was having difficulty explaining, he used the computer to both elucidate and validate his method.

Across all settings, negotiated interaction, sense of ownership and formal expression were not separately associated with learning but where all these aspects occurred together, settings were likely to be effective. Thus we can deduce that the interrelationship between task design and software environments, along with the availability of several computers, provided a fruitful basis from which good group practices could develop. In particular, the computers in our group settings allowed space for autonomous construction and validation of mathematical ideas. But not all group settings were as successful as the case study and where pupils disliked each other, unsuccessful competitive practices generally emerged. The availability of several computers exacerbates this tendency, as groups can split up and work towards separate goals – the pressure to be the first to construct an impressive computer product leading to a 'computer-centration' (see Hoyles, Healy and Pozzi, 1992). This pattern of working is likely to be unproductive and ineffective, militating against negotiation at the computer and synthesis of ideas away from the computer. Computer interaction tends to be dominated by pupil-directors, leaving others with little chance to express their mathematics formally. Additionally the group does not exploit opportunities to reflect upon diverse ideas, especially since written work is avoided or at best given scant attention.

In conclusion, our work suggests that the necessary condition for the success of a group setting is that pupils are sufficiently mature to manage themselves and their resources unimpeded by antagonism. Then, for mathematically based tasks, we argue that pupils should have the opportunity to formally express their ideas on the computer, develop a sense of ownership over these and reflect on them and those of others away from the computer. Within our study, this occurred when groups – through the activity of a pupil coordinator – exploited the task design by developing a framework of mutual interdependence. The implication one can draw is that activities are best seen as incorporating computer work rather than being computer-based, since it is the interweaving of on-computer and off-computer discussion which encourages pupils to make the necessary links between their own intuitions and classroom mathematics.

Notes

(1) Research project in conjunction with the University of Sussex funded by the InTER programme of the Economic and Social Research Council 1989–1991, Grant Number 203252006.
(2) A term taken from Lave (Lave, 1988), who highlights the importance of studying activity within its social context.
(3) In the Oracle study, this term was used to describe situations where pupils pursued individual goals, but at the same time and place as their neighbours (Galton, Simon and Croll, 1980).
(4) Popularity and knowledge status were derived from the questionnaires completed by *all* the pupils in the class.

■ References

Aronson, E., Bridgman, D.L. and Gellner, R. (1978) The effects of a co-operative classroom structure on student behaviour and attitude. In D. Bar-Tal and A. Saze (eds) *Social Psychology of Education: Theory and Practice*. New York: John Wiley.

Barbieri, S. and Light, P. (1992) Interaction, gender and performance on a computer-based problem-solving task. In H. Mandel, E. De Corte, S.N. Bennett and H.F. Friedrich (eds) (1992 in press) *Learning and Instruction. European Research in an International Context*. Oxford: Pergamon.

Bartolini-Bussi, M. (1991) Social interaction and mathematical knowledge. *Proceedings of the 15th PME Conference* Vol. 1, 1–16.

Cobb, P. (1990) Multiple perspective. In L.P. Steffe and T. Wood (eds) *Transforming Children's Mathematics Education: International Perspectives*. Barcombe, East Sussex: Falmer Press.

Eraut, M. and Hoyles, C. (1988) Groupwork with computers. *Journal of Computer Assisted Learning*, 5 (1), March, 12–24.

Forman, E. and Cazden, C. (1985) Exploring Vygotskian perspectives in education: the cognitive value of peer interaction. In J. Wertsch (ed.) *Culture, Communication and Cognition: Vygotskian Perspectives*. Cambridge: Cambridge University Press.

Galton, M., Simon, B. and Croll, P. (1980) *Inside the Primary Classroom*. London: Routledge and Kegan Paul.

Hoyles, C. and Sutherland, R. (1989) *Logo Mathematics in the Classroom*. London: Routledge.

Hoyles, C., Healy, L. and Pozzi, S. (1992) Interdependence and autonomy: aspects of groupwork with computers. In H. Mandel, E. De Corte, S.N. Bennett and H.F. Friedrich (eds) (1992 in press) *Learning and Instruction. European Research in an International Context*. Oxford: Pergamon.

Hoyles, C., Healy, L. and Sutherland, R. (1991) Patterns of discussion between pupil pairs in computer and non-computer environments. *Journal of Computer Assisted Learning*, 7, 210–228.

Hoyles, C. and Noss, R. (1992 in press) A pedagogy for mathematical microworlds. *Educational Studies in Mathematics*.

Johnson, D. and Johnson, R. (1975) *Joining Together: Group Theory and Group Skills*. New Jersey: Prentice-Hall.

Lave, J. (1988) *Cognition in Practice*. Cambridge: Cambridge University Press.

Perret-Clermont, A.N. (1980) *Social Interaction and Cognitive Development in Children*. London: Academic Press.

Slavin, R. (1983) *Cooperative Learning*. New York: Longman.

Sutherland, R. (1989) Providing a computer-based framework for algebraic thinking. *Educational Studies in Mathematics*, 20, 317–344.

Webb, N.M. (1989) Peer interaction and learning in small groups. *International Journal of Educational Research*, 13 (6), 21–39.

3.4

Gender Differences in Pupils' Reaction to Practical Work

P. Murphy

■ Introduction

The low number of girls studying physical science after the age of 13 has been the subject of world-wide concern since the 1970s. The change of status of science from an optional to a compulsory subject in the curriculum of many countries is an attempt to alter this situation. This can only be effective, however, if the reason for girls' avoidance of science can be identified and counteracted. Much is known about the differences in attitudes, experiences and achievements of girls and boys as they relate to science. What currently concerns educators is the way these factors interrelate to create gender differences.

Recent curriculum innovations in science typically include an increased emphasis on 'active' learning. To ensure that such innovations are beneficial to all pupils it is important to consider how gender differences operate in the practical context. Unfortunately, there is a confusion of messages coming from the literature about girls' and boys' attitudes to, and performance on, practical activities. Omerod (1981), for example, found that 'liking of practical work' was a significant discriminator for boys in all three science subject choices, in that it was an added incentive to study science. This was not true for girls. Yet recent studies looking at classroom intervention strategies have recommended practical work, or 'active' work, as a way of combating gender differences (see, for example, Hildebrand, 1989). Some national science surveys have shown large differences in favour of boys on practical tests (Kelly, 1981), others either no differences or a trend in favour of girls (Department of Education and Science, 1988). Girls' lack of confidence in practical contexts and fear of practical equipment has also been commented on in several studies.

To understand these apparent contradictions it is necessary first to explore the nature of gender differences and the factors that determine them. This [article] looks at some of the relevant international research in the area and examines the findings in the light of the results of science surveys carried out in the UK.

■ Gender differences in achievement – the international scene

Outside the UK the well-known survey programmes of educational achievement are those of the International Association for the Evaluation of Educational Achievement (IEA) and the National Assessment of Educational Progress (NAEP) in the United States. The first wave of surveys for these programmes took place in the late 1960s and early 1970s. The questions used were almost all multiple-choice and the majority were strongly content-based. The UK programme of the Assessment of Performance Unit (APU) started later (1980–4) and used quite different measures of achievement. The science surveys included: three practical tests, measuring skills, observation skills and practical investigations; tests of science content; and other process skill measures (Murphy and Gott, 1984).

The IEA survey found that on average boys scored considerably better than girls in science achievement tests in all 19 countries surveyed. A study by Kelly (1981) explored this finding further. The study focused on the 14-year-olds in the IEA sample from 14 developed countries. The IEA science tests were written tests which included measures of chemistry, biology and physics content; laboratory practice, that is, knowledge of apparatus and experimental procedures; and attitudes to science. The results again showed boys ahead of girls in every branch of science. The largest differences were for physics and the practical sub-test. The smallest difference was for biology, with chemistry intermediary. Girls in some countries did achieve higher scores than boys in others. However, the nature and magnitude of the gender differences were consistent across the various countries. The NAEP (1978) programme in the USA similarly found boys ahead of girls at ages 9, 13 and 17 on tests of physical science content, and only small differences in favour of boys in biology.

The cross-cultural uniformity of gender differences in science achievement suggests that cultural factors alone cannot account for them. Kelly (1981) identified the 'masculine' image of science, common to all countries, as a contributory factor. She argued that as girls and boys have learnt to respond to gender appropriate situations, then a 'masculine' science will alienate girls and discourage their engagement with it. Kelly's study also showed that internationally boys had a greater liking for, and interest in, science. There were considerable variations between countries, indicating to Kelly that cultural expectations did influence attitudes, particularly girls'. She did note, however, that boys achieved better in science than girls with equally favourable attitudes.

The UK APU science surveys of 11-, 13- and 15-year-olds included tests of both the skills and content of science. Only a small number of the test questions used were multiple-choice. The results of the surveys are given in Table 1, along with the related information from the IEA and NAEP surveys already discussed. In some cases there is a suffix to indicate the age at which the gender difference emerges. Where there is no suffix this indicates that the difference or similarity in scores occurred across the ages tested. The APU results demonstrate that gender differences in favour of boys increase as pupils progress through school.

Table 1 also refers to the results of another regional survey, the British Columbia Science Surveys (BCSS) (Hobbs *et al.*, 1979). This is included because, like the APU surveys, pupils' achievements on both the skills and content of science are assessed. However, like the IEA and NAEP surveys, the tests were in written not practical mode and the questions were largely multiple-choice. The British Columbia results show boys' superiority to be restricted to tests of physics content and of measurement skills.

The IEA results found that boys continued to outperform girls even when their curriculum backgrounds were similar. This was not the case in the APU surveys. When pupils with the *same* curriculum backgrounds are compared all performance gaps at age 15, except those for applying physics concepts and the practical test-making and interpreting observations, disappear. The results show that even when able girls continue to study physics they do not achieve the

Table 1   Some international survey results.

APU test	Results from APU	Results of other surveys		
		IEA	NAEP	BCSS
Use of graphs, tables and charts	$B_{15} > G_{15}$			$B = G$
Use of apparatus and measuring instruments	$B_{15} > G_{15}$	$B > G$		$B > G$
Observation	$G > B$			
Interpretation	$B_{13,15} > G_{13,15}$			$B = G$
Application of:				
Biology concepts	$B = G$	$B > G$	$B > G$	$B = G$
Physics concepts	$B > G$	$B > G$	$B > G$	$B > G$
Chemistry concepts	$B_{15} > G_{15}$	$B > G$	$B > G$	$B = G$
Planning investigations	$B = G$	$B > G$		$B = G$
Performing investigations	$B = G$			

B denotes boys' performance, G that of girls; B_x denotes the performance of boys aged x. The symbol '>' should be read as 'better than'.

same level as boys as a group. The gap in physics achievement between girls and boys established at age 11 increases with age (Johnson and Murphy, 1986), a finding also reported in the other survey programmes.

The 1984 APU survey looks in some detail at pupils' interests and attitudes to science. The results show a polarization in the interests of boys and girls across the ages. Girls' interests lie in biological and medical applications, boys' in physics and technological applications. The same polarization is evident in the pastimes and hobbies reported by the pupils. Girls also saw science as having little relevance to the jobs they might choose, unlike boys. Indeed, the proportion of girls selecting a job drops if they perceive a high science content in it. At age 15 a markedly higher proportion of girls than boys describe physics and chemistry as difficult. These findings replicate the IEA survey results.

At the time of the last APU surveys the second round of IEA testing was being carried out (1983–4). Some of the results for different countries have been published which provide further information about the nature of gender differences in science. For example, the second international science study (SISS) for Japan and the USA (Humrich, 1987) again showed 14-year-old boys ahead of girls for each science subject. However, Japanese girls outperformed boys and girls in the USA in physics whereas Japanese boys achieved lower biology scores than boys in the USA. Japanese scores were lower than US scores on tests of knowledge but substantially higher on tests of comprehension and application. The differences in levels of achievement for girls and boys between the two countries appear to reflect the different curricular emphases. For example, in Japan there is an early focus on the development of reasoning skills. Mathematics, which is particularly influential in physics learning, is a highly valued subject for all pupils. Biology, on the other hand, is accorded relatively low priority. Clearly the educational objectives of a culture can influence gender effects.

The SISS results for Israel (Tamir, 1989) showed a similar pattern of differences to the first IEA study. However, by age 17 the gender differences in biology and chemistry had disappeared for science majors and the difference in practical laboratory skills evident at age 14 was not present at age 10 or 17. The study found that girls specializing in physics did less well than boys in all science areas; whereas female chemists and biologists only did less well in physics. It is evident from these results that girls' learning in physics is an international problem.

National survey results have consistently shown boys ahead of girls on physics tests. Recent research in Thailand, however, provides an exception to these findings (Harding et al., 1988; Toh, 1991). Seven sets of tests were used in the study: three practical tests (of manipulative skills, problem solving, and observational skills) and four paper-and-pencil tests (on the links between practical work and scientific knowledge; sources of evidence; scientific knowledge; and scientific attitudes). The results show girls at age 16–18 performing at least as well as boys in physics and better than boys in chemistry. In laboratory tests girls outperformed boys in both physics and chemistry. The researchers posit several reasons for these uncharacteristic results, among them that science in Thailand is compulsory; the teaching approach is practical; chemical tasks have a 'feminine' image, as do some of the practical physics tasks; there are no differential expectations of pupils; and females participate in all levels and fields of employment.

It appears possible to alter girls' underachievement in physics by reconsidering the cultural expectations of girls and boys and how these are reflected in the organization and values underpinning the school curriculum. There are indications from each of the national surveys that, in certain circumstances, tasks in biology and chemistry can favour either boys or girls. To understand these results it is necessary to examine the effects of particular features of the tasks – for example, the degree of openness; the type of solution sought; and the manner of response expected.

The polarization of pupils' out-of-school experiences, as evidenced in their chosen hobbies, pastimes and choice of reading and television (see Johnson and Murphy, 1986, for a discussion), indicates some of the ways in which perceptions of gender-appropriate behaviours are constructed. Such perceptions lead girls and boys to develop the different interests and attitudes to science demonstrated in the surveys. These in turn affect how they engage with school science and their subsequent achievements in it. How pupils interact with science also depends on the image of science that is represented to them in their culture. The uniformity of gender differences across countries gives support to the contention that science has a masculine image in many cultures. The Thai results reinforce this.

Messages about girls' and boys' liking for and achievement in practical work remain unclear. The APU results indicate that girls are well able to handle a range of practical situations and indeed do better than boys in practical tests of observation. The Thai research also demonstrates girls' competence in practical work. The difficulties girls appear to have relate to more traditional curriculum approaches to experimental science. For girls the approach to practical work, the purpose it is seen to serve and the types of problem addressed by it all influence their performance.

The factors which influence gender effects operate throughout pupils' schooling and appear to be particularly critical for their learning in the intermediary school years (ages 13–14). We now take a closer look at the research results to try to establish potential sources of gender differences in science achievement.

■ Differences in experience

Kelly (1987) found little relationship between previous science-related experience, subject choice and achievement. The APU results paint a different picture. In the APU science surveys boys and girls at ages 11 and 13 perform at the same level on assessments of the use of apparatus and measuring instruments. These tasks were practical and pupils' performance was judged largely by their actions. When performance on individual instruments is considered, girls as a group do significantly less well than boys on certain ones. Boys are better able than girls to use hand lenses and stopclocks at all ages, and microscopes, forcemeters, ammeters and voltmeters at ages 13 and 15. Pupils in the surveys were asked what experience they had, out of school, of the various measuring instruments used. The results show that boys' performance is better than girls' on precisely those instruments of which they claim to have more experience. The different experiences of pupils affect not only the skills they develop but also their understanding of the situations and problems where their skills can be used appropriately. To plan effective classroom strategies it is necessary to focus on both the nature of the differential experiences and their consequences. For example, boys are better able to use ammeters and voltmeters yet they do not have experience of these outside of school. They do, however, play more with electrical toys and gadgets than girls. Such play allows boys to develop a 'feel' for the effects of electricity and how it can be controlled and manipulated. This is an essential prerequisite for understanding how to measure it. For girls to overcome this lack of experience they need to be faced with problems whose solutions, as they perceive them, require certain measurements to be taken. In this way girls will select instruments themselves and engage with them purposefully. This is a very different curriculum strategy to simply encouraging practice with instruments.

Another example of the influence of different pastimes relates to girls' well-documented lack of experience of tinkering and modelling activities. A common strategy employed to overcome this is to provide young children with Lego to play with. Teachers, however, are often discouraged by girls' apparent failure to engage with it. When boys play with Lego they do so in a purposeful way. Their play allows them to establish the link between their purposes and the potential of Lego to match them. It is this relationship that girls need a chance to explore. To facilitate this teachers have to identify the problems that girls find motivating in which Lego can serve a useful function. The same holds true for other areas of girls' inexperience. Erickson and Farkas (1987) investigated gender differences in 17-year-old students' responses to science tasks. They found that females only drew on their school experience whereas males were able to draw on a combination of formal experience linked to their everyday 'common-sense' knowledge. This gave the males an advantage in generating scientific explanations. They also report on females' negative responses to science tasks related to their lack of confidence and fear of handling practical equipment such as bunsen burners and electrical circuits. Girls' aversion to electrical matters is a well-established phenomenon which persists beyond schooling.

When faced with unfamiliar situations it is natural to feel uncertain about them. Girls' lack of certain experiences means that they approach some learning situations in science with diffidence and fear. In the APU surveys one assessment included open-ended practical problem-solving tasks set in a variety of contexts. More girls than boys react negatively to overtly scientific investigations. They express a low opinion of their scientific ability and feel generally unable to respond to such tasks. If the same problem is set in an everyday context, and such scientific equipment as measuring cylinders and beakers are replaced with measuring jugs and plastic cups, these same girls feel competent to tackle it (DES, 1989).

However, such a strategy has its drawbacks. Faced with an apparently everyday problem both girls and boys, quite reasonably, seek everyday solutions. Consequently they tend not to control variables or to collect quantified data. This suggests that the confidence of girls in practical work might be enhanced by working initially in everyday contexts but with problems that can only be solved using some degree of rigour. Hence there is a need, understood by the pupil, for scientific equipment.

■ Differences in ways of experiencing

Research into gender differences (Chodorov, 1978) has related the different patterns of nurturing that many girls and boys receive to the different values and view of relevance that they develop. These lead them to look at the world in different ways. As a result, children come to school with learning styles already developed and with an understanding of what is and is not appropriate for them. What they judge to be appropriate they tackle with confidence, what they consider alien they tend to avoid.

Although girls outperform boys overall on practical observation tasks, they do less well than boys on observation tasks which have a 'masculine' content – for instance, classifying a variety of different screws. The same situation was noted in pupils' performance on practical investigations. When offered a choice of investigations, girls who were competent problem-solvers rejected the one with an electrical content as they 'did not understand electricity'. This was in spite of assurances that the solution did not depend on any specific understanding of electricity. Boys, on the other hand, tended to reject the domestic-orientated tasks. They did so not because they regard the tasks as outside their domain of competence but rather outside science. They have a restricted view of the purposes that their knowledge can and should serve.

The results of the APU science surveys show that, irrespective of what criterion is being assessed, questions which involve such content as health, reproduction, nutrition and domestic situations are generally answered by more girls than boys. The girls also tend to achieve higher scores on these questions. In situations with a more overtly 'masculine' content – for example, building sites, racing tracks, or anything with an electrical content – the converse is true. Similar content effects are clear in other survey findings (see, for example, NAEP, SISS and BCSS). Alienation ultimately leads to underachievement as girls and boys fail to engage with certain learning opportunities.

Evidence available from classroom interaction studies indicates that the differences in the nature of the feedback that girls and boys receive about their classwork leads girls to have lower expectations of success and affects the way pupils interpret future experiences (Dweck *et al.*, 1978). For example, when girls and boys were presented with an investigation based on content they had already met in class the boys felt confident that they knew the answer, while the girls were inhibited by the belief that they should know the answer.

Girls and boys do appear to experience practical work differently. Randall (1987) looked at pupil–teacher interactions in workshops and laboratories. She found that girls had more contacts with the teacher and of longer duration than boys. However, the girls' contacts included many requests for help and encouragement about what to do next. Randall found that teachers, rather than building up girls' self-confidence, accepted their dependence on them and thus reinforced their feelings of helplessness. The combination of girls' timidity and boys' bravado leads to girls being marginalized in laboratories. Effects of this kind will lead to a real lack of skills in girls and to a substantial problem in future motivation.

■ Differences in problem perception

An outcome of children's different images of the world and their places in it is that the problems that girls and boys perceive are often very different given the same circumstances. Typically girls tend to value the circumstances that activities are presented in and consider that they give meaning to the task. They do not abstract issues from their context. Conversely, boys as a group do consider the issues in isolation and judge the content and context to be irrelevant. An example of this effect occurred when pupils designed model boats to go round the world and were investigating how much load they would support. Some of the girls were observed collecting watering cans, spoons and hairdryers. The teachers assumed that they had not 'understood' the problem. However, as the girls explained, if you are going around the world you need to consider the boat's stability in monsoons, whirlpools and gales – conditions they attempted to recreate.

In another situation pupils were investigating which material would keep them warmer when stranded on a mountainside. They were expected to compare how well the materials kept cans of hot water warm. Again, girls were seen to be doing things 'off task'. For example, they cut out prototype coats, dipped the materials in water and blew cold air through them. These girls took seriously the human dilemma presented. It therefore mattered how porous the material was to wind, how waterproof and whether indeed it was suitable for the making of a coat.

These examples of girls' solutions are often judged as failure either because their problems are not recognized or because they are not valued. Such responses to girls' perceptions of problems means that not only do they typically not receive feedback about their actions but any feedback they may receive will require them to deny the validity of their own experiences. This is a deeply alienating experience.

Another consequence of boys' and girls' different outlooks on the world is that they pay attention to different features of phenomena. A review of the APU item bank reveals that boys more often focus on mechanical and structural details, which reflects their greater involvement with modelling and handling mechanical gadgets both in and outside school. Girls attend to colours, textures, sounds and smells, data boys typically ignore. Consequently, girls always do better than boys, for example, on tasks concerned with chromatography irrespective of the scientific understanding demanded. It is easy to see how these differences go some way to explaining pupils' levels of achievement in physics and chemistry. Girls can deal with mechanical details just as boys are able, for example, to distinguish smells, but their different experiences result in alternative values and perceptions of relevance. When observing or interpreting phenomena pupils will deal with different selections of data unless teachers are alert to these effects.

Earlier mention was made of pupils designing model boats to go round the world. This task, along with one asking for the design of a new vehicle, was given to pupils aged from 8 to 15. The pupils' designs covered a wide range but there were striking differences between those of boys and girls. The majority of the boats designed by primary and lower secondary school boys were powerboats or battleships of some kind or other. The detail the boys included varied, but generally there was elaborate weaponry and next to no living facilities. Other features included detailed mechanisms for movement, navigation and waste disposal. The girls' boats were generally cruisers with a total absence of weaponry and a great deal of detail about living quarters and requirements, including food supplies and cleaning materials (notably absent from the boys' designs). Very few of the girls' designs included any mechanistic detail.

The choice of vehicle design and purpose also varied for girls and boys. Many boys chose army-type vehicles, 'secret' agent transport or sportscars. The girls mainly chose family cars for travelling, agricultural machines or children's play vehicles. The detail the pupils focused on was generally the same as before. Where boys and girls chose the same type of vehicle they still differed in their main design function. For example, the girls' pram design dealt with improving efficiency and safety. The boys' pram was computerized to allow infants to be transported without an accompanying adult.

These very different perceptions of the same problem reflect the way girls are encouraged to be concerned with everyday human needs. It was noteworthy that when the pupils had to focus on an aspect of their design the boys had few difficulties abandoning their initial design details. The girls, however, retained their details, which often made it difficult for them to pursue the teachers' focused investigation.

If a problem is perceived to be about human needs then it is not a simple matter to reject that perception and to focus on a more artificial concern, that is, the learning of a specific subject outcome. Yet this is commonly assumed to be unproblematic in science. Furthermore, an ability to do this may only reflect a limited and uncritical view of problems and problem-solving strategies. All pupils will benefit if some attempt is made to relate the focus to the larger problem and to discuss individual perspectives of relevance.

■ Pupil-friendly practical work?

Rennie (1987), reporting on a study of 13-year-olds in Western Australia, found that the relative inexperience of girls can be overcome in an activity-orientated style of science teaching. Similarly, the science curriculum of Thailand which appears to support girls' learning is described as activity-based, learner-centred, and focused on novel aspects of inquiry and discovery. Girls performed at a higher level than boys across the ages on the APU observation tests. The tasks used in these tests also tend to be novel and allow pupils to test out their own hypotheses before reaching a conclusion. It was found that girls' performance was significantly higher than boys' on such tasks, but particularly so when the tasks were both active and open. Furthermore, the nature of the task can override the content effects discussed earlier. For example, an observation task involving various electrical components in a circuit was popular with boys but not with girls. Boys' performance was also significantly higher than girls. Yet when the performance of girls who liked the type of task, in spite of the content, was compared with boys who similarly liked the task there were no gender differences in performance. These findings lend some support to the view of what a gender-inclusive science curriculum might be like.

It is often concluded from discussions of gender differences that girls need to address problems that reflect their social concerns. Yet the observation tasks described are not untypical science examples. It is important here to distinguish between accessibility and conditions for learning generally. The tasks used in the APU study are accessible because they allow pupils to formulate and test their own hypotheses. It is up to the pupil to accept or reject them on the basis of their own data. It is the pupils' views of relevance that prevail. However, in learning situations, if the new knowledge pupils acquire is to be linked to their existing knowledge then problems do need to be set in the context of human experience and needs. These provide the framework for girls to make sense of the problems posed and the motivation and purpose they need to continue with the complexity of learning in science. Of course, the same should obtain for boys.

■ Summary

Practical work can play a crucial role in combating gender differences. As the discussion has pointed out, it matters what type of practical work is encouraged. Strategies for changing the aims and nature of science learning to take account of gender differences must be applied thoughtfully. There is little point in attempting to extend the accessibility of science if in the process other groups of pupils become alienated. Classroom strategies will have to take account of boys' and girls' present preferred styles of working and interests as well as providing opportunities for them to reflect critically on them. Many of the tactics that need to be employed to facilitate girls' scientific learning are merely examples of good teaching practice anyway – for example, setting tasks that allow all pupils to express their interests and understandings in a manner that is appropriate for them. However, it follows that tasks of this kind must be fairly broad and general, therefore that pupils will perceive different problems and different solutions within them. Of course, this has always been the classroom reality but the intention now is to make it an explicit and understood aspect of teaching practice. Consequently, there will be a need to adopt a more flexible approach to curriculum planning. A much broader view of the potential learning outcomes in any one lesson or module will have to be accepted by teachers. Similarly, consideration will have to be given to the numerous learning pathways that pupils can follow in acquiring particular scientific understanding. This is necessary because teachers have to focus pupils' learning and need to be aware of when and how to do this in a way that enables pupils to see the point of the focus and thus continue with their learning. The teacher's role will be even more demanding than at present if such strategies are adopted. On the plus side, they will be rewarded by the commitment and achievements of pupils who hitherto have found it difficult to make sense of science. These pupils at the moment not only remain scientifically illiterate but spend a lot of their time in school pursuing purposelessness and meaningless activities.

■ References

Chodorov, N. (1978) *The Reproduction of Mothering*. Berkeley: University of California Press.

Department of Education and Science (1988) *Science in Schools Age 15. Review Report*. London: HMSO.

Department of Education and Science (1989) *Science in Schools Age 13. Review Report*. London: DES.

Dweck, C. *et al.* (1978) Sex differences in learned helplessness: the contingencies of evaluative feedback in the classroom. *Developmental Psychology*, 14, 268–76.

Erickson, G. and Farkas, S. (1987) Prior experience: a factor which may contribute to male dominance in science. Contributions to the Fourth GASAT Conference 2, Michigan, USA.

Harding, J., Hildebrand, G. and Klainin, S. (1988) International concerns in gender and science/technology. *Educational Review*, 40, 185–93.

Hildebrand, G. (1989) Creating a gender-inclusive science education. *Australian Science Teachers Journal*, 35, 7–16.

Hobbs, E.D., Bolt, W.B., Erickson, G., Quelch, T.P. and Sieban, B.A. (1979) *British Columbia Science Assessment (1978)*, General Report, Vol. 1. Victoria, BC: Ministry of Education.

Humrich, E. (1987) Girls in science: US and Japan. Contributions to the Fourth GASAT Conference 1, Michigan, USA.

Johnson, S. and Murphy, P. (1986) *Girls and Physics*. London: DES.

Kelly, A. (1981) Sex differences in science achievement. In A. Kelly (ed.) *The Missing Half*. Manchester: Manchester University Press.

Kelly, A. (1987) Does that train set matter? Scientific hobbies and science achievement and choice. Contributions to the Fourth GASAT Conference 1, Michigan, USA.

Murphy, P. and Gott, R. (1984) *Science Assessment Framework Age 13 and 15*, Science Report for Teachers: 2, ASE.

National Assessment of Educational Progress (1978) *Science Achievement in the Schools*. A summary of results from the 1976–77 National Assessment of Science, Washington Education Commission of The States, Washington, DC.

Omerod, M.B. (1981) Factors differentially affecting the science subject preferences, choices and attitudes of girls and boys. In A. Kelly (ed.) *The Missing Half*. Manchester: Manchester University Press.

Randall, G.J. (1987) Gender differences in pupil–teacher interactions in workshops and laboratories. In G. Weiner and M. Arnot (eds) *Gender under Scrutiny*. Milton Keynes: Open University Press.

Rennie, L.J. (1987) Out of school science: are gender differences related to subsequent attitudes and achievement in science? Contributions to the Fourth GASAT Conference 2. Michigan, USA.

Tamir, P. (1989) Gender differences in science education as revealed by the Second International Science Study. Contributions to the Fifth GASAT Conference. Haifa, Israel.

Toh, K.-A. (1991) Factors affecting success in science investigations. In B.E. Woolnough (ed.) *Practical Science*. Milton Keynes: Open University Press, pp. 89–100.

PART 4

Learning Technology

Discriminating between Levels of Capability

Assessment of Performance Unit

[In this second extract from the APU report, evidence on the various aspects of capability is given. All pupils were given a holistic mark to reflect their overall capability. In this extract the focus is on finding out what aspects of capability contribute to the holistic mark. These 'aspects' are elaborated into three kinds of categories: procedures, communications and concepts, which are further broken down into headline categories: procedural qualities (reflective, appraisal and active); communication qualities (the complexity, clarity, confidence and skill of communication); conceptual qualities (understanding and using materials, energy systems, aesthetics and people). Further, procedural capability can have reflective and active qualities and appraisal qualities that link the other two together.]

■ The constituents of good performance

[. . .] we saw in the last section how the distribution of holistic marks for different tests begins to indicate important differences in the nature of those tests as well as telling us important things about the nature of pupil performance. However, whilst this debate remains at the holistic level, it is difficult to get into enough detail to unravel the crucial indicators that discriminate between mediocre and good performance. This is the issue that we now intend to address, for we are interested to discover how patterns of performance in the three key areas of the assessment framework [procedures, communication and concepts] alter at different holistic levels. What qualities discriminate between a good and a poor design and technologist?

Our assessment framework describes capability in design and technology in terms of three principal components:

157

- *processes* of design and technological activity (procedures)
- *communicative*/modelling facility (communication)
- *conceptual* understanding (concepts)

and these categories are the basis on which we have constructed this more detailed examination. We constructed *performance profiles* (based on the three categories) which allowed us to examine the relationships between the levels achieved in these critical areas of capability and the 'holistic' mark of performance.

■ Constructing performance profiles

We took all pupils with holistic scores of 4 and 5 (our highest overall performers) and contrasted their performance profiles with pupils who had scored 2 holistically. We chose these levels because if 2 typically represents a large number of pupils that might be described as average/poor, and 4/5 represents the relative few that are our best performers, then the difference between them should be of great importance for teachers. The differences in the profiles ought to pin-point where *in particular* the 4/5ers are better than the 2ers and (by extension) where teachers might concentrate their energies if they wish to elevate the performance of the average group.

[. . .]

The graph [in Figure 1] provides an overall picture of the survey by comparing the profiles of *all* pupils with high holistic scores and *all* pupils with average holistic scores from *all* our tests. The profiles are very similar, and form a parallel pattern suggesting that:

- overall, both high and average holistic performance comprise similar ingredients in similar proportions – high performers generally display more of everything;
- proportionately, high performers are slightly more active than average performers;
- in terms of procedures, our tests would seem to have generated slightly more reflective than active responses;
- generally, pupils are confident in their communication;
- pupils demonstrate least understanding of energy systems in our activities and most understanding of people.

This global picture is interesting, but it obscures more than it reveals, for when we look at the profiles for different test structures, we see some important differences that provide very interesting insights into the nature of capability.

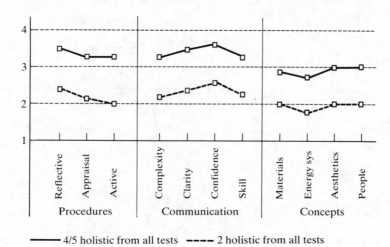

——— 4/5 holistic from all tests ‒ ‒ ‒ 2 holistic from all tests

Figure 1 Performance profiles for pupils with high holistic scores and average holistic scores from all tests.

[. . .] we use this global picture as the basic yardstick against which to measure the differences across tests. Interestingly, the Early Ideas test has a profile that conforms very closely to that of the whole sample. Marginal differences suggest that this Early Ideas test shows greater discrimination (between 2ers and 4/5ers) for active procedures, for clarity in communication and for energy systems in conceptual terms. But the differences are marginal as can be seen in Figure 2.

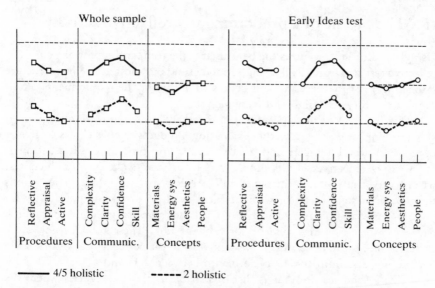

——— 4/5 holistic ‒ ‒ ‒ ‒ 2 holistic

Figure 2 Performance profiles for all tests and Early Ideas tests.

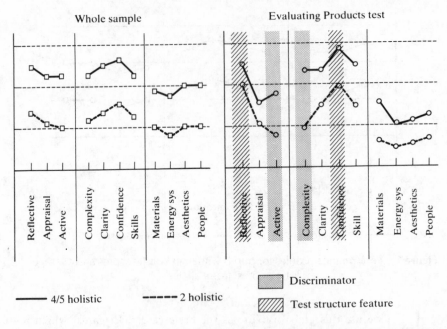

Whole sample   Evaluating Products test

—— 4/5 holistic   ----- 2 holistic

Discriminator

Test structure feature

Figure 3 A comparison of the performance profiles for all tests and Evaluating Products tests.

■ The profile of a predominantly reflective activity

Using the 'all tests' profile as the benchmark, the comparison between our most reflective test structure (Evaluating Products) and our most active test structure (Modelling) is very revealing and begins to highlight the requirements of all-round capability in design and technology.

Two features of these profiles are immediately distinguishable [Figure 3]. First, as one would expect, the Evaluating Products profile shows a greater dependency (for both 2ers and 4/5ers) on *reflective* procedures. Evaluating is essentially a reflective activity. Moreover, [. . .] the Evaluating Products test have a very tight – and therefore supportive – structure that carries pupils through from start to finish without them needing to take responsibility for deciding what to do next. In such a test it is not surprising to find that a high *confidence* measure is also evident for both groups. These two categories are therefore interesting in terms of how the tests affect performance, but they are not the big discriminators of performance between 2ers and 4/5ers.

The second feature that is distinguishable is those categories where the two lines diverge, for these are the major discriminators between average and good performance. Clearly, they show up as *active* procedures and *complexity* in communicating and at first sight the former might seem odd. However, the message would appear to be that high-level performance in evaluating involves more than the passive observation of qualities. It does involve (and require) such high-level reflection, but it is also strongly linked to the ability actively to generate and develop ideas for improving the product/system that is being evaluated. This has important repercussions for ensuring a balance of activities.

The second major discriminator is related to the first. The 2ers (who were relatively reluctant to move into active mode and consciously grapple with the consequences of their judgements by improving the product) were also less able to come to terms with the complexity of the task. A feature of the Evaluating Products test was that much of the communication was based – of necessity – on the written word, and it was principally in the area of re-design for improvement that it was appropriate for pupils to use other (more graphic) communication styles. We believe that it is not just a coincidence that those who were more able to move into active mode and explore the possibilities for re-design were also those who could handle more of the complexity of the task. It is precisely as our model in [Article 2.2] outlines; the more detail with which we can express our thoughts, the more detailed those thoughts can be. Active engagement in re-design of the product brings into sharp focus the complexities of the product and – by extension – enhances one's ability to evaluate it.

■ The profile of a predominantly active activity

The most contrasted test to Evaluating Products (in terms of the active/reflective balance of the activity) is the Modelling test, where not only were pupils consciously engaged in developing solutions, but moreover they were doing so with the use of real materials and often engaging in discussion and collaboration. The discrimination profiles for 2ers and 4/5ers are inevitably different, therefore, but again provide an interesting perspective on the nature of capability [see Figure 4].

One first obvious point to make is that once again *active* procedures are a clear discriminator, with the 4/5ers rating proportionately very much higher than the 2ers. This is perhaps less surprising in an essentially active test, but it is nonetheless important to note. Interestingly, however, confidence in communicating also emerges (though less clearly) as a factor, and again this is probably related to the issue of active procedures. Those who are less prepared to get involved in active design and development may be the ones who are less confident in their ability to explain their ideas in material form. The Modelling tests are procedurally the loosest and least supportive of all our tests and if pupils are essentially unsure or unconfident in exploring and developing ideas, we might expect it to show up in the confidence measure and the active capability measure.

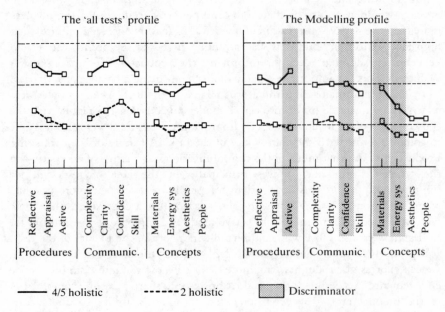

The 'all tests' profile   The Modelling profile

——— 4/5 holistic   ------2 holistic   ▨ Discriminator

Figure 4 The performance profiles for all tests and the Modelling tests.

By contrast, the fact that (in conceptual areas) materials scored relatively highly for both groups is not surprising, and can be put down to a Modelling test structure effect, in which we deliberately allowed (indeed encouraged) pupils to explore their ideas through materials. The 4/5ers emerge as considerably better at it, however, as they are with exploring energy systems.

The question of how conceptual demand varies through the test is highlighted by the comparison between the Modelling test profile and that for Starting Points [Figure 5], which is the test in which pupils are involved in trying to identify and tie down a specific task (or starting point) for a project, from a broad context. It is immediately observable that whilst the Modelling tests made considerable demands on pupils' understanding of materials and energy systems, the Starting Points test highlights the crucial understanding of *people*. If one is trying to identify needs within a context, and specify with some clarity what a good starting point might be for a project, then understanding people (the users) is clearly going to be important. The profile confirms this commonsense view and makes it the biggest conceptual discriminator of performance between 2ers and 4/5ers [Figure 5]. At the other extreme, it is interesting to note that understanding and using energy systems did not score highly for either group, which is not surprising in that those sorts of understandings are typically more central to activity in the later developmental stages of an activity.

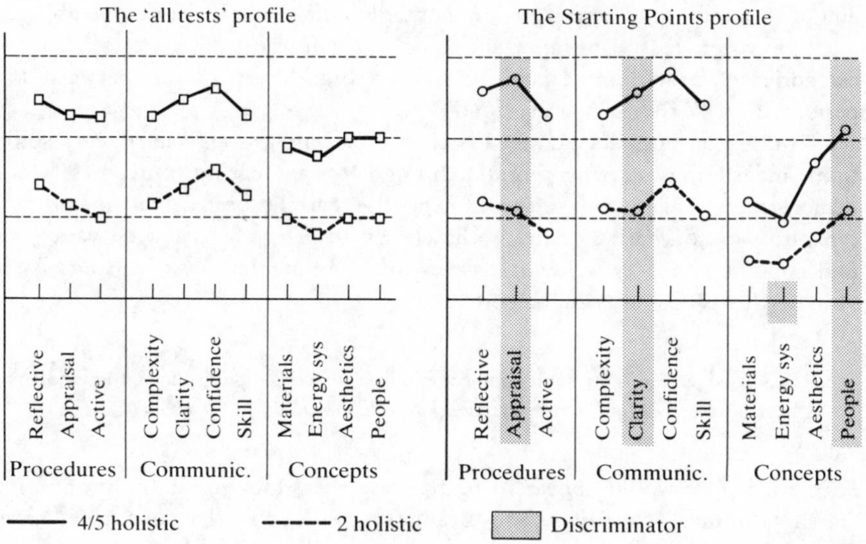
The 'all tests' profile
The Starting Points profile
Reflective
Appraisal
Active
Complexity
Clarity
Confidence
Skill
Materials
Energy sys
Aesthetics
People
Procedures
Communic.
Concepts
4/5 holistic
2 holistic
Discriminator

Figure 5 A comparison of the performance tests for all tests and the Starting Points tests.

In the other categories, it is interesting that again confidence emerges as a high scorer for both groups [Figure 5] and is consistent with the position in the Evaluating Products tests. As in those tests, the structure of the Starting Points tests is tight and supportive, giving clear directions as to what is expected, and this has clearly raised the confidence level of all pupils. Clarity in expressing ideas emerges, however, as (marginally) the greater discriminator and is probably related to the fact that appraisal (in the procedural categories) discriminates best. Appraisal must be seen here in terms of the judgements pupils are making about the relative importance of needs and opportunities within the context, and in terms of how good their starting point might be in satisfying them.

Summarizing the discriminators

Taken as a whole, there are not many of the categories within our assessment framework that do not emerge as important discriminators of capability in one or more of the tests. Some, however, are clearly more crucial than others.

The ability to engage *actively* in the development of ideas and proposals for products/systems is clearly a central plank of capability. Whether the focus of work is on identifying a starting point, on exploring early ideas, on modelling a solution, or on evaluating an outcome, the result is the same. Performance in any of these stages of activity is undoubtedly enhanced by this active ability.

However, it is also important to note that in most of the tests, all three categories of procedural capability show a big differentiation between the performance of the 2ers and 4/5ers.

In the conceptual areas, the discriminators vary as the focus of work shifts from identifying starting points through to evaluating outcomes. This demonstrates – again as one would expect – that the best overall measure of capability in conceptual terms is the ability to *make use of* a wide variety of understandings *as they become appropriate* to the immediate task in hand. As it says in the N.C. Interim Report:

> 'Knowledge here is a resource inseparable from practical action, not a commodity to be stockpiled before action can begin.' (*DES/WO, 1988, para 2.12*)

Furthermore, it would appear to be the case that, looking at the profiles for Starting Points, Modelling, and Evaluating Products, the conceptual areas show less discrimination overall (between 2ers and 4/5ers) than either the procedural or the communication categories.

In terms of communication, the biggest discriminators again vary from test to test: clarity is important in identifying starting points, confidence is more important when modelling outcomes, and complexity when evaluating them. Having said this, however, *all tests* show relatively large differentiation between the two groups of pupils for *all* the communication headings, so we are really seeing a difference of emphasis between tests rather than major differences of focus. It is a somewhat obvious truth that communication is the means by which the other measures are made observable, and the shifts of emphasis from test to test are probably most important in what they tell us about how pupils are able to respond in different stages of a task. 2ers, for example, find it relatively hard to be clear about a starting point, because it is a hard thing to do to pin down the detail of a task from a nebulous context. Moreover, they are less confident in resolving a task, and less able to penetrate the complexity of potential outcomes. These are overall observations on the nature of capability as much as measures of communication.

■ The interaction of elements

A major weakness of this type of analysis, however, is that it is too easy to see the categories and the qualities as independent – which we know for a fact that they are not. The individual marks that make up the profile are interacting with each other, and we have attempted in the above analysis to suggest some of the major interactions. We initiated this analysis of profiles having seen how the patterns of holistic performance varied and seeking to explain them in more detail. The profiles are important in helping in that explanation, but [. . .] the complex integration of capabilities affects performance.

Using the procedural categories [. . .] we saw that, in most cases, integrated capability (linking active and reflective through appraisal) results in a high holistic score. Similarly a low score for these procedural qualities usually results in a low holistic mark. While these extremes seem fairly well defined, the procedural requirements for achieving a high holistic score actually fall on a continuum:

- it is *necessary* to show high levels of *active* or *reflective* ability but this is not sufficient to guarantee high holistic performance;
- if you show both *active* and *reflective* ability you are *more likely* to score high holistically;
- if you can integrate *active* and *reflective* ability with appraisal you are *most likely* to score high holistically.

However, while this does appear to be generally true, it is not universally so, and in approximately 7% of cases the formula breaks down and fails to explain the holistic mark. In these cases, some of which were genuinely odd pieces of work, the disparity is probably only explicable in terms of the interaction of elements, and not just the procedural ones. These difficulties raise important questions as to how far any aggregation model could cope with what is clearly such an integrated set of capabilities.

■ Detailed performance descriptors

[Elsewhere in the report] we describe the structure for the three levels of the assessment framework: *holistic* judgements, headline *categories*, and individual *descriptors*. So far in this [article], the profiles have been composed from the headline category judgements, with the holistic used just to fix the profile as a 2 or a 4/5. However, we can also use the individual descriptors to help us to construct a more detailed picture within the profile, thereby giving a yet more detailed picture of what constitutes good or average/poor performance.

The following lists detail the descriptors of the ways in which the high performers worked where there was a significant difference between them and the average performers. We have only listed those descriptors when the difference between the 2ers and 4/5ers performance is significant at the 1% level. [. . .] For clarity, we have grouped the descriptors into the three broad groups of reflective, active and appraisal.

☐   **Reflective abilities supporting high holistic performance**

- pupils show a high level of understanding of the *central purpose* of the task (i.e. what it must do);
- they identify both user and product issues *of their own accord*, as well as when asked to do so as part of the activity;
- they *focus development* on both product and user issues;
- they raise issues that *go beyond* the introductory information given in the video and the workbook task box;
- they are *prolific* in raising issues, which are *wide-ranging*, *detailed*, *explained* and *justified*;
- in their proposals, they *work on* most of the issues they raise, including the three they identify as most important.

☐   **Active abilities supporting high holistic performance**

- pupils *tackle* the central purpose of their task with a high level of understanding (having decided what it must do, they do something about it);
- when they make proposals for
  - how things will work
  - how they will be used
  - what they will look and feel like
  they do so with a *high level of understanding*;
- their proposals focus on both *user and manufacturing* issues;
- they brainstorm *lots* of starting points, are *prolific* in making proposals and *make significant progress* towards a solution;
- they show significant evidence of *to-ing* and *fro-ing* between thought and action;
- they show evidence of being *divergent* thinkers and introduce new ideas in their later work;
- they are *systematic and rigorous*, pushing at 'good' ideas remorselessly;
- their ideas approach a *working reality* taking account of both users and manufacturers.

☐   **Appraisal that supports high holistic performance**

- they appraise the *central purpose* of their task with a high level of understanding (having done something, they see how it stands up to what they said needed to be done);

- when they identify *strengths and weaknesses* in their proposals for
  - how things work
  - how they are used
  - what they will look and feel like

  they do so *of their own accord* as well as when the test demands it, and they do it with a high level of understanding;
- they *justify* their judgements in terms of
  - economics
  - aesthetics
  - technical success;
- they *identify tensions* between strengths and weaknesses;
- they are *prolific* in making judgements and they focus on both *user and manufacturing* issues.

Taken together, the above lists provide a rich picture of positive working methods. While we would not want these to be seen in a prescriptive way, we do believe they indicate strengths in ways of working that could beneficially be developed in lower achievers.

■ Performance profiles and individual pupils

We have attempted, in this [article], to indicate some of the major ways in which holistic performance on our tests can be explained. We have done this through a comparison between those who have performed well on our tests and those who have done relatively poorly. As such this analysis begins to highlight some of the critical aspects of capability and (equally importantly) it helps teachers to see how they might support the poor performers and help them to improve. [. . .]

■ Reference

Department of Education and Science and the Welsh Office (DES/WO) (1988) National Curriculum Design and Technology Working Group Interim Report. London: DES/ WO.

4.2

Using Science in Technology Projects

G. Job

■ Science and technology as curriculum activities

□ Curriculum models and stereotypes

[. . .] Curriculum developers [have a] responsibility to ensure that curricula do not reinforce some of the potent oversimplistic stereotypes of science and technology [found in the world], made more acute by the fact that very few teachers have ever themselves been practising scientists or technologists, and very few science and technology teachers have any significant training in history, sociology, or epistemology. The requirement for an accurate awareness element of science and technology courses has to be seen in the long-standing English tradition that the essence of a school subject is the teaching of the subject, from which learning about the subject should emerge (McCulloch *et al.*, 1985). Courses that, for example, directly teach pupils about technology with relatively little emphasis on developing their subject skills are almost unheard of in England, unlike the USA, for example, where such courses are not only acceptable and well resourced, but also have as a consequence a well-developed theory and methodology.

Any course that attempts to give students a realistic awareness of the relationship between science and technology is faced by the inevitably limited experience and understanding of students, perhaps particularly in science. Two distinct approaches seem to be common. One is to relate some aspect of doable and understandable 'school science' to a simple explanation of a feature of a technological artifact or system – e.g. using simple practical and theory work on Bernouilli's principle to 'explain' the movement of a wind generator rotor.

This may root the scientific principle in a real, high-tech context but does nothing to illuminate the interplay between scientific and technological principles needed to make the generator a practical possibility. Another common approach is to extrapolate from a simple school-based application of a scientific principle to a large-scale, real-world analogue – e.g. from a laboratory working model telegraph to satellite communication. Here the technological principle may be illustrated, but the science involved in the real-world example is irrelevant to its school model.

[. . .]

☐ 'Doing' science and technology

A dominant feature of English curriculum development over the last 30 years has been the conviction that, at least in practically related subjects, learning in the subject is best achieved through 'doing' the subject. One of the most overworked slogans has been the probably mythical, and allegedly Chinese, proverb: 'I hear and forget, I see and remember, I do and understand.' Although there is certainly some research support for the efficacy of this general approach, most associated developments in science and technology have been predominantly based in pragmatic enthusiasm (McCulloch *et al.*, 1985). (It is interesting that only two recent major science curriculum projects in the UK have had a strong base in learning theory: the 'Science 5–13' project (Schools Council, 1974), explicitly based on Piaget's work, and the 'Integrated Science Project' (SCISP, 1973), which rigorously applied Gagne's model of learning. Both have been very influential on curriculum debate but have had minimal take-up in schools.) There is no doubt that this emphasis on 'learning by doing' has transformed for the better the practice of many teachers of science and technology related subjects; but these transformations may owe as much to rekindled enthusiasm and the Hawthorne effect as they do to curriculum theory and research.

In English science teaching, this practical approach to science, with its classroom model of 'the child as scientist' (Driver, 1983), was galvanized by the Nuffield Foundation's science projects of the 1960s, and is exemplified in two aspects of national science policy. In the 1970s, central government established an Assessment of Performance Unit (APU) to monitor children's performance in key subjects of the curriculum. The assessment categories used for science (APU, 1987a) were:

(1) Using symbolic representations
(2) Using apparatus and measuring instruments
(3) Observation
(4) Interpretation and application
    (a) of given and everyday information

(b)   of taught science concepts

(5)   Planning investigations

(6)   Performing investigations.

This model of what is worth assessing in science education clearly views 'practical investigation' as the culmination of school science activity. The 'whole investigation' is the opportunity for the student to bring together practical skills and theoretical understanding in order to operationalize a scientific problem, select and control appropriate variables, design and execute relevant experiments, and analyse and evaluate the results. This research-focused model of science has been carried forward into the National Curriculum for science, despite government concerns that learning of scientific fact and theory may be somewhat undervalued. The National Curriculum (DES/WO, 1989) defines two major elements ('Profile Components') for science for all pupils from the ages of 5 to 16:

(1)   Exploration of science

(2)   Knowledge and understanding of science
      with the former carrying between 30% and 50% of assessment weighting.

A similar progress can be traced in English technology education over the last 30 years, although the detailed history is immensely more complex than for science, with educational and political demands from the diverse traditions of the crafts, engineering, the sciences, and design, among others, for a legitimate say in the nature of technology in the curriculum (Cross and McCormick, 1986; McCulloch *et al.*, 1985). Despite this complexity, a common theme has been the stress that the essence of the technology curriculum should be 'doing technology', based on sometimes conflicting models of what it is that technologists do. As with science, the formulation of the Assessment of Performance Unit (APU) has been highly influential (APU, 1981, 1987b). This focuses on the development of students' capabilities in three areas:

(1)   *Procedural* – the ability to select and apply appropriate design and technology procedures in identifying tasks, investigating them, generating and developing solutions, and evaluating them.

(2)   *Conceptual* – the ability to demonstrate and exploit platforms of conceptual knowledge (e.g. of materials, energy, control) in resolving the task that is set.

(3)   *Practical* – the ability to manifest (e.g. through sketches, diagrams, notes, and models) the depth and development of thinking.

(Although the APU identified four value areas relevant to technological activity – technical, economic, aesthetic, and moral – these have not so far been pursued in assessment surveys.)

This emphasis upon the task as the heart of school technology activity has now found a statutory base in the National Curriculum for Technology (DES/WO, 1990). This identifies two 'Profile Components':

(1)   Design and technology capability[1]
(2)   Information technology capability.

The former is defined in terms of four attainment targets:

(1)   Identifying needs and opportunities
(2)   Generating a design
(3)   Planning and making
(4)   Evaluating.

The accompanying programmes of study and criteria for assessment elaborate this task-based model of technology. Skills of drawing, modelling, and making are important only in so far as they contribute to successful task execution. Similarly knowledge about specific materials, energy resources, or control systems is only necessary if it resources the task. Knowledge and skills are seen as resource bases to be drawn upon as required in exploring and developing possible designs to meet a need.

☐   **Science investigations, technology tasks, and 'capability'**

Comparison of these two process-driven modes of curriculum activity – the scientific investigation and the technological task – is instructive. They both reflect a concern to develop capability in students, to get students involved in making decisions about what is worth doing, about how best to do it, and about what needs to be learned to do it (Layton, 1984; RSA, 1984). In these respects, they both fit into that category of activities that Black and Harrison have called 'Task–Action–Capability' which, they argue, epitomize all 'technological' activity (Black and Harrison, 1985) – an all-embracing use of the word technology which further illustrates the etymological confusion in the English curriculum debate.

Certainly the scientific investigation and the technological task share many common features, and are perhaps best seen as defining a spectrum of activity rather than distinct activities. When students – and adults – really become involved in them, they are activities driven by curiosity about the natural or human-made world, and a realization that that curiosity can be satisfied by purposeful combination of practical and mental action. At the science end of the spectrum, the curiosity may be predominantly on how and why something happens and at the technology end on how to make it happen differently or better. Experience, particularly with younger children and their teachers, suggests that from the point of view of children's developing interest and action these categoric distinctions are largely artificial.

What paradigm of 'real science' underlies this emphasis on the science investigation? Its emphasis on systematic observation, gathering of evidence, and the cautious drawing of generalizations suggests a model that is empiricist, inductive, and clarificatory and that owes more to Bacon, Hume, and Braithwaite than to Popper or Kuhn. In practice, it is probably not a bad model for giving a view of the work of the ordinary, journeyman scientist, but it does not fit that well with the more heroic model of competing theories and scientific revolutions that is often used to give spice to the teaching of scientific theory.

This curricular emphasis on the systematic science investigation can, however, make a valuable contribution to the development of good technological task activity. It can bring rigour and system to the evaluation phase of technological activity, where alternative solutions are to be fairly evaluated or where optimization of the desired solution is being attempted. Evidence from public examinations, from APU surveys, and from discussion with teachers shows that this aspect of task work does not come at all naturally to children. Evaluation of otherwise enterprising and thoughtfully executed tasks is effectively confined to what has been called 'up a bit, down a bit' evaluation. One reason for this may be related to the limited experience of many teachers in these aspects. The hard fact remains that very few teachers of science or technology have been practising scientists or technologists; and it appears that the systematic control of variables in the context of real-life situations is something that teachers find hard to translate from theory to practice.

■ Applying scientific knowledge in technological activity

☐ The problem of transfer

[. . .] Conventional wisdom suggests that students' technological capability will be enriched if they have knowledge and understanding of a wide range of scientific concepts that they can bring to bear on their technological tasks. Science is an essential 'resource base' for technology. The wider and deeper the students' science knowledge, the more flexible and sophisticated their response to a technological challenge. If they have studied enzymes as well as catalysts, they have the opportunity to explore biotechnological as well as chemical solutions to a technological challenge. Only with an understanding of basic electric circuit theory can they be expected to use microelectronic circuits effectively. An understanding of issues of transportation surely requires a study and some understanding of basic kinematics and dynamics. So goes the conventional curriculum argument. Hence the attempts to co-ordinate or even integrate the school science and technology curricula. And when these attempts fail, or are not even attempted, we see the depressing features of bad science inappropriately taught in technology – hard sums in workshops – or technology badly executed in science – shoddy model bridges, inedible cakes, corrosive cosmetics in laboratories. Detailed curriculum co-ordination in schools between subject areas is notoriously difficult to achieve. How important in practice is such detailed content co-ordination between science and technology? How transferable to technology activity is scientific knowledge, and how impoverished is students' technology without it?

A number of writers have suggested that technological activity involves a significantly distinct 'way of knowing' from that involved in scientific activity (Fores and Rey, 1986). The latter involves 'propositional' or 'explicit' knowledge epitomized in academic debate, the language and mode of knowing of 'homo sapiens'. By contrast, technological activity involves the 'transactional' or 'tacit' knowledge of the doer, the inventor, the maker, 'homo faber'. (This distinction is caricatured in the old joke: the engineer doesn't know why his bridge stays up; the scientist knows why his fell down.) Any teacher who has undertaken technological project work with children or adolescents knows that accurate knowledge and understanding of scientific principles is rarely a prerequisite to practical invention and progress to an effective solution. The explanation of the solution may require a sophisticated treatment of those principles; the development of the solution appears often not to.

It can, for example, appear bewildering to a teacher that a student can develop, make, and refine a device that scans the bar-codes on tinned goods, decodes them, and produces a voice-synthesized output; and yet the same student can have difficulties in explaining the functions and modes of operation of the fuse and earth (ground) lead in an electrical circuit.[2] However, analysis of the procedure followed by the student in the project activity suggests that, despite its complexity, the points at which the student had to understand (in the scientific sense) an electrical or electronic concept in order to make effective progress were actually very few; and that the concepts were basic and largely qualitative (e.g. the ideas of a continuous circuit, of voltage as push and current as flow, and of a voltage pulse). Much of the remaining electronic activity involved the creative interrelating of a number of known or researched functional blocks. In such an activity, the intervention of analysis of the block or the linkage of blocks can appear to be detrimental to success. One key to success seems to lie in having intelligent access to resource bases of relevant functional blocks, with sufficient scientific understanding to recognize issues of matching and linkage.

This way of using scientific principles poses major challenges for the science curriculum. It is an uncomfortable fact that, certainly for school students, 'wrong' science is as good as, and often better than, 'correct science' in this respect. Aristotelean physics, the caloric theory of heat, fluid theories of electricity can all deliver the technological goods (or 'good-enoughs'). To deliver 20th century scientific ideas clearly, plausibly, and economically is the technological curriculum challenge to science.

☐ An agenda for research

There appears to have been little systematic research into how children in practice learn and develop competence in technology, in particular how they undertake problem-solving activities and utilize their existing knowledge and skills in science. Comparable recent research into the development of children's scientific skills and understanding, e.g. the work of the CLISP project in England (Driver, 1988), has signalled the importance of working from the reality of children's existing conceptual frameworks, with major implications for effective pedagogy and curriculum development. As technology education as a component of the entitlement curriculum develops, and as it becomes increasingly urgent to lay a firm basis within the entitlement curriculum for the continued development of advanced technologies, the highest priority needs to be given to developing such a research base. This task should be in itself a paradigm for that blend of creative problem solving and analytical rigour which a symbiosis of science and technology can produce.

Notes

(1) The curriculum terminology surrounding technology in the UK is involuted, to say the least. The current position, broadly speaking, is that Craft, Design, Technology (CDT), a reasonably well established subject area, is a subset of Design and Technology, a multi-subject area, which is a subset of Technology, which may also include Computing and Information Technology. This semantic structure owes more to educational, political, and academic power struggles than it does to epistemology, curriculum theory, or common sense.

(2) These examples are based on the author's discussions with 14- to 16-year-old students following separate science and technology courses leading to GCSE examinations, and with their science and technology teachers.

■ References

Assessment of Performance Unit (APU) (1981) *Understanding Design and Technology*. London: HMSO.

Assessment of Performance Unit (APU) (1987a) *Science at Age 13: A Review of APU Survey Findings 1980–84*. London: HMSO.

Assessment of Performance Unit (APU) (1987b) *Design and Technological Activity: A Framework for Assessment.* London: HMSO.

Black, P. and Harrison, G. (1985) *In Place of Confusion.* London: Nuffield-Chelsea Curriculum Trust. [See Article 2.1.]

Cross, A. and McCormick, R. (eds) (1986) *Technology in Schools.* Milton Keynes: The Open University Press.

Department of Education and Science and the Welsh Office (DES/WO) (1989) *Science in the National Curriculum.* London: HMSO.

Department of Education and Science and the Welsh Office (DES/WO) (1990) *Technology in the National Curriculum.* London: HMSO.

Driver, R. (1983) *The Pupil as Scientist?* Milton Keynes: The Open University Press.

Driver, R. (1988) Restructuring the science curriculum: some implications of studies on learning for curriculum development. In D. Layton (ed.) *Innovations in Science and Technology Education*, Vol. II, pp. 59–83. Paris: UNESCO.

Fores, M.J. and Rey, L. (1986) Technik: the relevance of a missing concept. In A. Cross and R. McCormick (eds) *Technology in Schools.* Milton Keynes: The Open University.

Layton, D. (ed.) (1984) *The Alternative Road: The Rehabilitation of the Practical.* Leeds: University of Leeds.

McCulloch, G., Jenkins, E. and Layton, D. (1985) *Technological Revolution? The Politics of School Science and Technology in England and Wales since 1945.* Lewes: The Falmer Press.

Royal Society for Arts (RSA) (1984) Manifesto of education for capability. *RSA Newsletter*, Spring.

Schools Council Integrated Science Project (SCISP) (1973) *Patterns: Teacher's Handbook.* London: Longmans.

Schools Council (1974) *With Objectives in Mind.* 5–13 Science Project. London: Macdonalds.

4.3

Learning Design and Technology in Primary Schools

A. Anning

■ Introduction

I have a long established interest in enhancing the status of 'the intelligent hand' and it was this aspect of technology education that appealed to me. I set about monitoring the first year of technology in 12 primary schools in two local education authorities (LEAs). Gill Kicks, a researcher with a teaching and town planning background, collected most of the data, and this is a first attempt to make some sense of what we found. In the autumn term we visited the 12 schools to talk to the headteachers, technology co-ordinators and teachers of Years 1 and 3. They were still reeling under the stress of the requirements to teach to the English, Mathematics and Science Orders. A sense of 'overload' was a recurrent complaint. As one teacher succinctly put it, 'It's getting ridiculous'.

■ Children's learning in design and technology

We returned to four classrooms, two Year 1 and two Year 3 age groups, to observe children working on activities defined by their teachers as design and technology. We wanted to explore how the teachers translated their strategic models of design and technology into operational models. We also wanted to collect empirical evidence of young children's capability to set against levels of attainment for Key Stages 1 and 2. We used a mixture of field notes, tape recordings, photographs, photocopies of children's work and stimulus material, and video recordings, to collect detailed data.

Time and time again teachers ask for research findings to be presented in a way that relates to their language and everyday classroom concerns. I have chosen to illustrate some of the many general issues arising from our analysis of the data through the specifics of four classroom incidents. The four exemplars have been selected to illustrate particular features of capability and related implications for teaching and learning: young children's thinking through modelling, investigating the properties of materials, learner-directed activities and teacher-directed tasks.

☐   Young children's thinking through modelling

Exemplar 1, Key Stage 1, Designing an exercise area for the school hamster (James 5.5 and Stephen 5.4)   The task was in response to a real need. The hamster, constipated and unhappy, had been taken to the vet and declared to be in need of exercise. The vet's diagnosis was discussed at the start of the day class session.

The two boys were encouraged to draw their ideas first. The researcher's suggestion that they might 'scribble' down a few preliminary ideas met with looks of astonishment and disbelief. Instead the two boys embarked on an elaborate, detailed and highly imaginative drawing. They drew a table and chairs with (no doubt healthy!) breakfast cereals, a bed with pillows and duvet, a jacuzzi for the hamster to relax in, a pool for it to drink from and a train set for it to play with. Their only concession to hamster need was to define his colour preferences as pink and yellow.

Finally they began to draw in cage-like elements on the design – horizontal bars for the sides 'so that he can poke his little head out' and a door 'because he does need to go out for exercise'. When it was suggested that they might write labels for parts of the design to help them when they made the model – the rudiments of an annotated drawing – they politely declined 'because we're not very good writers'. It became increasingly clear that they intended their drawing for the hamster, 'He's only a little baby hamster and he can't write . . .'. At this point the children were making no links at all between their detailed 2D drawing and the possibility of translating their ideas into a 3D model. 'We don't want to make a model, do we?' was the dominant child, James' firm response.

The next day the teacher re-focused the boys' attention on making a 3D model. She helped them to identify and draw up a list of materials they would need – coloured paper (they chose gold, silver and black – a concession to the concept of a metal cage?), coloured chalk (James had been waiting for his opportunity to get hold of these!), a box, carpet and wallpaper pieces, mirror, special paper to look through (for windows), string, wood, paint and brushes, a hole punch (for air holes), and a knife to cut window shapes (to be controlled by the teacher). It was clear from their talk and gestures that the boys were now imaging with materials clearly in mind.

On task they began by cutting the flaps off the top of a large box and carefully painting the outside, pausing only to stick small squares of gold and silver paper on the surface. Did these represent windows or were they a reference to metal? James then set Stephen the task of measuring and cutting a piece of string which they attempted to glue across the top of the box onto the still wet surface. The researcher suggested they cut slits in the box to fix the string. They explained that the string was for the hamster to run along. Had James seen the television advertisement where the squirrel runs the gauntlet of the washing line to reach the nuts? James painted the string, but then began to worry that 'I've made it a bit slippy'. He sent Stephen off to find something soft for the hamster to land on in the event of a fall. Instead Stephen returned with some wood offcuts and the boys began to make a diving board. When they could not attach it to the top of the box, they simply converted it into a bed. James then turned his attention to cutting some steps out of card for the hamster to climb up to the string. Card steps proved to be too difficult to make. Ever the pragmatist, James decided that the hamster could use 'his sharp little claws' instead. However, by chance he discovered some thin strips of balsa amongst the wood offcuts and immediately began to concentrate on sticking them with wood glue up the inside of the box to make a ladder.

This long session of designing as they made occupied the boys for several hours. They returned sporadically to the task of mechanically re-painting the surface of the box, as if this gave them time for reflection.

Discussion   One of the issues illustrated within this exemplar is the use of drawing in promoting young children's designerly thinking. Many of the teachers we interviewed spoke of misgivings about the appropriateness of requiring children at Key Stage 1 to draw out a design. An infant teacher said, 'I don't think they need to draw and plan it at five. I think they want to make it first and possibly record and draw about it afterwards. They don't know what it looks like before they make it. To see it before you've done it, that's hard.' Even at Key Stage 2, teachers spoke about the difficulties of getting children to produce realistic design drawings. 'The designs they make at the beginning, however fantastic they look, very often the end product isn't like that at all. I think that what actually happens is that they are re-designing as they go along all the time, as they are making things . . . They are always over-ambitious in the drawings.'

The intellectual demands we are making of children in asking them to represent a 3D model in 2D form would tax many adults. The ability to visualize objects in diagrammatic form and translate these images into line drawings, with all the attendant complexities of perspective, scale and overlap, is a particularly sophisticated, taught convention.

We need to *teach* children a range of drawing conventions, the equivalent of teaching them a range of writing registers. This can start at Key Stage 1 by sharing talk about different kinds of diagrams and their appropriateness for different purposes — cross-sectional drawings, architects' plans, exploded diagrams, botanical drawings, annotated drawings, rough sketches. These are simply the equivalent resources in graphic terms to the range of reading material — fiction, non-fiction, reference, catalogues, comics, etc. — provided in many classrooms.

Children should be encouraged to draw in different ways for different purposes. James and Stephen were imbued with the notion of a drawing as a finished product, in this case a gift for the hamster, not as a tool for thinking. They had no experience, within school conventions, of the way designers use sketches to clarify their emergent ideas. We saw other examples of confusion about the purpose of design drawings. A group of Key Stage 2 children had drawn designs for decorating pump bags. When they actually began to cut out the shapes they had drawn in felt they had great technical difficulties. They realized that they needed to simplify their designs drastically but, concerned that their original drawings should be seen as 'wrong', they surreptitiously rubbed out sections to produce 'correct' versions for the teacher.

The exemplar also illustrates the dilemma of children 'imaging' without information. The children had no obvious resources to support their imaging of a design for a hamster exercise area — no access to a real cage or commercially produced hamster exercise aid, nor even catalogues with photographs of equipment for pets. Their thinking was therefore constrained by images of artefacts from their own world.

In other instances, the 'over-ambitious' design drawings we saw children produce at Key Stage 2 had no underpinning of (a) a conceptual knowledge base of the properties of materials to be used, or (b) the production techniques required to translate images into outcomes. In most cases, teachers did not focus children's attention on the need to plan with these constraints in mind. Here the teacher-directed listing of materials, identified by the children as what they needed to make their model, served the useful functions of (a) getting the children to image with materials in mind, and (b) training them to use advance organizers in translating design intentions into achievable outcomes.

A further issue is a general lack of clarity we found about design intentions. In this example, the hamster was never given the opportunity to try out his gymnasium. It was not clear if the boys ever believed that this was a real possibility or if it mattered to them whether they were making a model, a prototype or an artefact with a real purpose. Both the drawing and making tasks were productive in all kinds of ways for the children's learning. We observed many excellent examples of children working in kinesthetic mode in what appeared to be an *ad hoc* way but which provided systematic opportunities for them to practise and refine techniques of handling materials and tools while 'designing as they did'. However, lack of clarity about the purpose of tasks, both in the teachers' and the children's minds, could result in aimless and unproductive model-making sessions. As one teacher said, 'It's making children aware of their design capability that is important. Whilst children are simply making "houses" out of cornflakes packets, there is little sense of purpose in that. It isn't to investigate the properties of a house (or the properties of card) or anything of that nature. It's to produce a house that can stand on a shelf and look nice in front of some written work or something.'

☐ Investigating the properties of materials

Exemplar 2, Key Stage 1, Experimenting with paper structures (Sarah, Tracey, John, Martin, Sean, 5-year-olds)   The teacher introduced the task to the group of children by demonstrating ways of strengthening paper – rolling, folding, layering, etc. The instructions were to build upwards by experimenting with paper. Resources provided were scissors, sellotape, strips of paper ready cut, approximately 9 in by 1 in.

Left to their own devices, initially all five children started to make paper chains. They tested for 'strength' by blowing the chains and judged the strongest to be the one that moved least! The teacher intervened to refocus them by demonstrating coiling and folding paper again and reminded them that they were supposed to be building upwards with the paper. The children began to work independently on a range of interesting structures [Figure 1].

Sarah built a series of platforms using coiled paper for columns and flat strips of paper as cross-pieces. Tracey working alongside began with a similar structure as a base and tried to build up looped paper strips on top of her base. When it toppled sideways, she ingeniously used a circle of card sellotaped to the base platform to stabilize the whole structure.

John worked with great concentration on rings of paper with doubled-over strips inserted inside like a lantern. His was the tallest and sturdiest structure. But an indication that he was in fact drawing on a mental model of

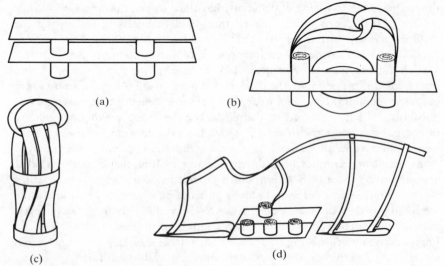

(a)   (b)

(c)   (d)

Figure 1   (a) Sarah's paper structures; (b) Tracey's paper structures; (c) John's paper structures; (d) Martin's paper structures.

paper lanterns for his construction was that he added a 'handle' to the top ring. Sean continued to experiment with paper chains. Using the kind of lateral thinking typical of streetwise five-year-olds, he taped the final ring of his chain to the back of his chair so that it dangled to the floor and announced that he had made his stand and it was the tallest! Martin made an elaborate device of coiled paper and strips [Figures 2 and 3], but soon became frustrated when he could not attach side supports to make it stand up. The researcher demonstrated to him the concept of triangular support frames using folded paper strips but he did not have the manipulative skills to use the ruler and scoring technique she showed him, and he abandoned the task.

Discussion   It seems essential for children to have opportunities to explore the qualities of materials and techniques for fixing and joining them. In the NCC document, *Issues in Design and Technology* (NCC, 1991, p. 10) a distinction is made between *resource tasks* (designed to help pupils acquire the knowledge, understanding and skills necessary for design and technology capability) and *capability tasks* (designed to provide pupils with opportunities to demonstrate capability). Here investigation of the properties of paper and ways of using it to build structures (resource tasks) underpinned a subsequent set of activities designed to build and test the weight-carrying strength of 'bridges' made of paper (capability tasks). In the other Key Stage 1 class, we observed children moving on from freely experimenting with ways of fixing card together – with hole punchers, paper clips, staples, butterfly fasteners, plant wire – to the task of making Christmas cards with moving parts. This task was resourced by examples of books and cards with moving parts provided by the teacher. The combination of the experience of experimenting with materials and skills for fixing and joining them, and teacher-led discussions about real cards and pop-up books, led to successful and innovative designs. Analysing the component parts of tasks and planning systematically to build up a knowledge base and set of practical skills over a half-term would ensure progression in children's learning.

One way of managing exploratory work would be to set up a workshop area where the resources would be structured over a period of weeks to focus on strands within the Programmes of Study – for example, joining materials and components in a simple way. If activities involving making vehicles were planned, the workshop area would be set up with construction kits, and box modelling materials with containers of wheels and axles sorted and categorized into shapes and sizes. Adjacent pinboards would (a) have outlines to indicate where tools should be stored when not in use, thus training children to keep the working surfaces tidy and safe, and (b) skill cards and prompts to demonstrate different methods of making wheels and axles, fixing chassis, etc. Children can be given the freedom to experiment without the pressure of a task which requires an end product. Class or group discussions can be used to draw attention to interesting experiments. Children can be encouraged to keep their own records of what they have achieved over a half-term in notebook form – much like a designer's pad. Exploratory work then becomes much more purposeful for children.

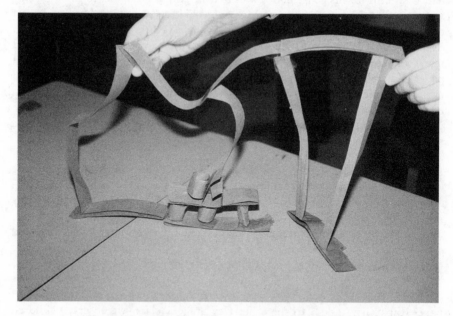

Figure 2   Martin's structure in need of support.

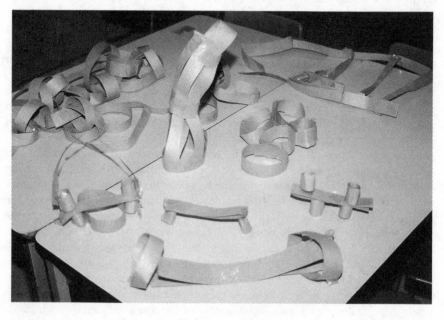

Figure 3   All the paper structures.

☐   **Learner-directed activities**

Exemplar 3, Key Stage 2, Making a model of a traction engine (Susan, Shelley, Ruth, 7-year-olds)   The girls had come across instructions to make a traction engine model in a library book: 'I'd been wanting to make it for ages 'cos it looked right good'. They asked their teacher if they could make the model. She encouraged them to list the materials and equipment they needed. In a workshop area set up in a corner of the classroom, they assembled boxes, corrugated card, PVA glue, a glue stick, a large pair of scissors, pea-sticks, string and sellotape.

They began by casting around for containers to draw round to make four large and four small circles for the wheels. Following the diagrams in the instructions, they made tyres by cutting strips of corrugated card, measuring the length by eye, glueing them to the circles and chopping off any overlap. They completely missed a crucial strengthening device of cotton reel inserts for the wheels. Prompted by the sight of an eggbox and card tube in a container nearby, Susan turned her attention to another section of the instructions, scheduled for much later, and began to make the smokestack: 'Don't know what it's for. Might be for the smoke, but it's got that thing on top.'

Returning to the task that afternoon, the children decided to use a block of wood for the main body of the engine. They took turns to struggle, without a bench hook or clamp, to saw the block to size, again gauged by eye. They

Figure 4   The completed engine.

attached the smokestack using the technique, learned from a sculptor (recently artist in residence in the school), of a set of card struts folded at angles of 90 degrees and glued to the body of the engine and the upright chimney [Figure 4].

Their choice of the wooden block caused a series of technical difficulties. The steering and axle mechanisms of the illustrated model involved making holes in the main body through which to feed the pea-sticks. LEA policy restricted the use of a drill in school until a drill-stand had been delivered. The children resorted to drilling holes with the sharp end of a large pair of scissors and a small screwdriver to attach the front axle! In the end, they attached the surface paraphenalia of a steering mechanism – a cotton reel and length of string attached to the front axle – but never got it operating. They fixed a smart little card canopy, as illustrated in the instructions, with pea-sticks attached to the wooden block by copious lengths of sellotape. They coated the finished model with white emulsion paint – again a technique learned from the sculptor – and decorated it with powder paint using unsuitably large brushes.

In all the girls had spent sessions spread over five days on the model and were proud of it. 'We worked hard, didn't we?' When asked about possible improvements, they acknowledged that they knew how they could make the steering wheel work. 'We've got to put a pencil through there, but we can't be bothered. We *can* make a hole in it, but it takes too long.'

Discussion   The children made a laudable attempt at following quite complicated diagrammatic instructions but, without a teacher on tap to guide them at decision points, they were frustrated by difficulties which could have been avoided. We observed that children often selected materials because of their outside dimensions rather than their workable properties, and were seduced by the novelty of unfamiliar or exotic materials. For these children a brand new block of wood looked enticing. If an adult had drawn the children's attention to the internal workings of the model as represented in the diagram, which showed a hollow box for this component, they might have been dissuaded from using the wooden block. But in this classroom, as in many, the teacher was managing a whole class of lively Year 3 children on a range of curriculum activities, and was only able to pay sporadic attention to the workshop area.

The teachers we interviewed had adopted a range of strategies to make time to teach and monitor technology. Some teachers timetabled sessions when they targeted the technology activity for their 'quality time'. Some set up whole days when all the children did practical activities. Many of these were 'holding' activities in which the children were well trained to operate independently – role play, construction, sand – while the teacher focused her attention on small groups. Others involved parent helpers.

The teachers were perplexed by the difficulty of getting children to *evaluate* their work, particularly when the outcome was the result of personal choice and commitment. These girls were quite open and confident about evaluating their product, but in general teachers commented that they found children reluctant to make negative comments about their work. 'We ask the child when they've finished a model or a piece of work, are you pleased with it or would you like to change it, but they usually say "Well, I like it as it is".' Teachers felt that it was unrealistic to expect a young child to re-do work – 'it might put them off' – and negative feedback from the teacher at the end of a task was seen as potentially unproductive. We observed that sensitive ongoing evaluations fed back into the task as it evolved, provided insights which improved the quality of subsequent work. But in order to do this, teachers had to pay regular and careful attention to what was happening while the children were on task. Since practical work has traditionally been regarded as something the children can be left to get on with, this requires a major shift of attitude for primary teachers.

☐　**Teacher-directed tasks**

Exemplar 4, Key Stage 2, Making a wooden noughts and crosses game (groups of six Year 3 children)　The task was for each child to mark out a standard grid pattern on squared paper, transfer the grid design by pressing drawing pins through the paper sellotaped onto a ready-cut square of wood, to drill holes for wooden pegs, to sand the surface of the wood, and thus to make a simple board game. The objectives of the task were to give the children practice in copying diagrams, to teach them how to operate a drill, and for them to make a Christmas gift to take home.

　　With these specific objectives in mind, the sessions were characterized by the tight control of a well-trained parent helper. She insisted on high standards. The children were expected to work methodically through the stages and help each other where one of a pair was struggling. The drills were small with pistol grip handles, attractive to look at, but quite awkward for small children to grip at table height, and no concession was made to left-handers – they had to turn the handle towards them. In each pair one child held the block steady while the other drilled. The incidental talk was productive – about the number of turns required and the amount of pressure they needed to exert on the drill, the heat of the friction, the age rings and smell of the newly cut wood. Though the pace was brisk, the children enjoyed the sessions, and the quality of the completed board games was uniformly high.

Discussion　Training and supervising children in the use of unfamiliar tools was a constant source of anxiety for the teachers. 'If they have an accident with paint or something on their clothing, parents expect that, but if they go home with a cut from a saw I think you'd have a lot of justifying to do if you weren't there.' Teachers also worried about the classroom furniture! 'It's not only inflicting injuries on themselves or other people, but . . . the furniture, tables and chairs and the floor, you know. You might find yourself ending up with a pile of firewood.' The kind of supervision illustrated by this exemplar may seem regimental. It certainly goes against the grain for many primary teachers who would argue for the benefits of an informal teaching and learning style such as that illustrated in Exemplar 3. But the objectives of the task were achieved. Everybody learned how to use a hand drill safely.

We observed recurrent problems with equipment. We tend to forget how difficult some of the manipulative skills implicit in design and technology activities are. Often children have ambitions to make things which the limitations of their hand/eye skills simply cannot match. We saw many children getting quite literally tied up in knots of sellotape – masking tape seemed a lot easier tô handle. We also found that small bench hooks, particularly when they were not clamped to surfaces, were very difficult for children to use. Pushing against the hook with one hand, while sawing with the other, left pieces of wood flailing about wildly. Saws were equally tricky, as were a range of drills we saw children operating. Snips were useful, but most of the younger children needed two hands to cut with them. We saw collections of tools on smart display boards in primary classrooms which simply did not function properly in children's hands. We need research which monitors accurately the development of hand/eye manipulative skills in young children so that tools can be designed to be more user friendly for young children.

■ Conclusion

Despite the inauspicious context into which Key Stages 1 and 2 Technology was introduced in 1990 we were impressed by the capability children were already displaying and by the commitment of their teachers to experiment with new areas of knowledge and skills.

Difficult though it was for the teachers to absorb the model of design and technology outlined in the Order (the iterative nature of designing and doing) our observations of children confirmed that this was the way in which they worked most productively. They shifted from talking, to manipulating materials, to drawing, to copying others, to seeking help from more knowledgeable peers or adults, to trying things out, to modifying and evaluating as they worked through a task.

The teachers were also learning as they went along. They recognized the need to resource children's thinking by providing opportunities for them to build up a knowledge base to underpin their design decisions. They were beginning to understand when and how they should demonstrate practical skills to children and to rethink the management of their time to allow for this. They recognized that the organization of workshop space, storage of tools, ways of working collaboratively, and generating a climate of 'critical friendship' for evaluation, all needed to be 'taught' rather than 'caught'.

We could see the potential in design and technology for children to demonstrate 'creative and practical capabilities'. Technology offers opportunities to pay real rather than rhetorical attention to practical work in primary classrooms.

Acknowledgements

My thanks to Gill Kicks and David Layton for comments on an earlier draft.

■ Reference

National Curriculum Council (1991) *Issues in Design and Technology, Key Stages 1 to 4, Teachers' Notes*. York: National Curriculum Council.

4.4

The Importance of Graphic Modelling in Design Activity

S.W. Garner

[The one-year research project funded by the National Society for Education in Art and Design (NSEAD) and described here by Steve Garner resulted in some illuminating findings from the case studies. The research aimed to promote a re-evaluation of drawing within design activity. This project led to a larger programme of research named ROCOCO (Remote Co-operation and Communication) which involved Garner and which was led by Dr S. Scrivener of the Department of Computer Studies at Loughborough University. This focused on the communicational requirements of designers who are cooperating remotely – that is, using computer technology to link design teams who may be some distance apart. The resulting research, involving verbal and non-verbal communication, is currently being written up.]

■ Pump-priming by the NSEAD

The NSEAD-funded one-year programme of research set out to examine the functions of drawing for a wide range of professionals engaged in three-dimensional design. It followed the popular method of case-study analysis but, because of the variety of information this technique generates, the opportunities for examining and testing particular hypotheses were limited. Nevertheless, it was a method particularly appropriate to that investigative programme as it provided a very broad foundation upon which later studies could be based. The findings from the research have been presented in a number of papers.[1,2] In these papers, drawing is presented as a major influence on the development of designerly thought, and comparisons are made between the investigation of drawing and the studies of natural language by researchers such as Bruner[3] and Barnes.[4]

The case studies consisted of 20 examples of designers and included, among others, architects, sculptors, engineers, theatre designers and crafts-people. Some of these held teaching posts in their subject. The analysis involved looking for commonalities of opinion and was based on transcripts of each interview. These transcripts were bound together and submitted to the NSEAD at the conclusion of the work. In one of these transcripts, Dick Powell[5] refers to the capacity of drawing to allow a designer to 'converse with themselves'. Such conversation may involve asking the right questions, constructing the right structures and providing conjecture. Assimilation of information may be more important than communication and drawing may provide a means of achieving this. A number of case studies referred to a use for drawings in 'turning over fresh information' or 'consolidating a theme'. This 'homing-in' extends the contribution of drawing from a problem-solving to a problem-finding strategy. Client-based meetings were cited as one situation where drawing had an important function in the formulation and definition of requirements.

Exploratory drawing would appear to increase the potential of discovering or seeing new arrangements or directions in old information. Obviously the individuals' creativity will have some bearing on this but there is some evidence to suggest that the development of creativity is in some way encouraged by the implementation of graphic strategies. The technical skills of representation are clearly going to influence the ability to portray what is intended but the development of representational drawing skills demands that the draftsperson looks and 'sees' with greater insight. That is to say, there is a relationship between drawing and visual literacy. Thoroughness of seeing and the development of critical abilities are similarly interrelated. One of the research subjects (a furniture designer) promoted the activity of observational drawing as a means of exploring, understanding, remembering and, particularly, critically judging, and in this way influencing the quality of concept, detailing and proportion. There was some agreement among the case studies for this capacity as Claire Webber, a painter, commented:

'I draw to help me understand. It's rarely used to express myself. It's to do with learning about what you are looking at and being surprised.'[6]

There appears to be a deliberate and structured approach in the exploitation of drawing for responding to external and internal stimuli. Drawing appears to be used for the deliberate provocation of responses rather than as a passive medium for externalizing, in an intuitive manner, such stimuli. Responses need not be restricted by practicality. A free flow of ideas may be essential and distinct from the evaluative strategies employed in the design process. Roy Axe, Director of Concept Engineering at Austin Rover at the time of the research, sees a long-term future for the notorious marker-pen rendering of the automotive industry simply because he can see no other means yet of capturing the essential 'caricature' of the designer's intention.[7] Perhaps the research work into 'computer-aided vehicle styling' currently being undertaken by Michael Tovey at Coventry Polytechnic will change this. The term 'spirit' may be viewed as synonymous with caricature as highlighted by another subject, this time an architect, in referring to one of the great designers of the 20th century:

> 'Alvar Aalto did very, very sketchy, embrionic, schematic drawings which were purposely ethereal because he was trying to catch what you could only call the spirit or the essence of the job. He did not wish to compromise solutions by seizing on form too quickly.'[8]

Somewhere between the analysis of the problem and the conscious exploration of potential resolutions lies a cloudy perceptual domain within which designers refer to sources of motivation or inspiration that result from quick sketches they have made. Certain sketches would seem to be produced within which reside sufficient ambiguity for the mind to see a variety of subsequent moves. Thus a creative analysis is begun that appears to display some congruity with Tovey's dual-processing model of activity within the mind.[9] Case-study evidence seems to support this by reference to the importance of drawings which possess 'flexibility', that is, drawings which can be interpreted in a number of ways. Central to this issue is the deliberate reduction of pre-conceived 'meaning' without the sacrifice of 'feeling'. Closely associated with this is 'serendipity' or happy chance. This was so commonly mentioned by subjects that one is led to believe that happy chance can be conjured up at will. Perhaps the immediacy of drawing assists this process. Similarly humour has been proposed as a catalyst in this context whereby drawing may act as a trigger. In the same way that re-interpretation (or misinterpretation) of information may lead to unusual perceptions and subsequently lead to laughter, so interpretation of graphic information may trigger creative insights. The use of the word 'play' occurs frequently – and a little apologetically – within the transcripts, as if there should be no requirement for such apparently unfocused activity. On the contrary, play can be an immensely focused activity (observe young children with a favourite toy). Economists, scientists, ecologists as well as members of the design professions appear to be quite happy with the notion of 'playing' with ideas as an essential stage to creative idea development.

Underlying the exploitation of drawing as an exploratory tool is a considerably more basic foundation. This is the sheer enjoyment in the activity. It appears that the subjects of the research would draw during design activity whether it presented an advantage or not! Even in homes and offices the numerous doodles in telephone pads would appear to indicate that drawing can be organizational, creative and involuntary. It is difficult to isolate the functional requirements to draw from this inner motivation but a great many designers appear to be unable to stop themselves from drawing as they talk, listen or create. Whether drawing ability promotes design practice or vice versa is unclear, but the majority of those interviewed regularly made time to maintain their graphic abilities through recreational sketching or painting. In one discussion, Alan Williams of DCA design consultants explored the relationship between recreational sketching and the exploitation of drawing within the pressured, commercial world of product design. His belief is that designers require a constant interest or inquisitiveness in the world around them and that sketching, with its requirements for looking and thinking, is an important way to maintain this.[10] One subject referred to his National Service in this context:

> 'When I went into the army I continued drawing. I've got sketch books full of drawings simply because it was a way of coming to terms with the world around me; new landscapes, new situations, new people. It seemed to me a way of making contact in a very real way.'[11]

The relationship between drawing and designing is further articulated by Norman McNally of Glasgow School of Art:

> 'If you cannot report on what exists, i.e. you do not have an investigative vision of the world around you, then you can hardly be expected to report on what does not exist – things that you are pulling out of your head. Objective drawing constantly informs conceptual drawing.'[12]

Drawings produced during design activity may be intended to possess no more than fleeting value and yet they can often display characteristics that make them very precious to the drawer – and the viewer. A blend of serendipity, skill, speed, economy, pleasure, pain, anger and humour can often produce a sketch of more interest to people than the finished product – whether that be a building, a domestic product, a piece of sculpture or a painting. In fact the roughness of a sketch would appear to be an important characteristic of some types of design drawing. A very detailed 'conceptual' sketch might stifle ambiguity and might fix early thoughts which could be improved upon. Subjects were often apologetic for the quality of some sketch work shown but this is likely to result from the general assumption within our society that drawings are meant to convey information. If we suspend the notion that drawings are communicative devices and search for their meaning in developmental strategies of the mind then we may come closer to an appreciation of their role.

The role of drawing as a modelling device within an exploratory strategy is clearly, then, not limited to small patches of application. It lies at the very heart of the human search for understanding. Exploration has been presented as a conglomeration of interrelated activities, some revolving around problem analysis and inquisitiveness, others around creativity or discovery and still others that are concerned with making visible the products of such exploration.

The area within the design process between exploration and idea development is a very grey one. Rarely does one get the opportunity to complete research activity thoroughly before manipulating the information in a response to identified requirements. In fact, a case could be made for the importance of creatively examining the breadth of a problem and the possible responses to it before a systematic research process is completed. Thus drawing strategies that aim to explore problems, manipulate information and visualize responses have no clear divisions between them. Designing is not a linear process. Its iterative nature is well accepted and this results in differing requirements for drawing at any one stage of the process. To compound this issue, skilled practitioners, as represented in the case-study subjects, are able to produce drawings which have multiple functions and, more importantly, functions which take place apparently simultaneously. To take an extreme example, a sketch which may have been made to externalize a private and incomplete notion may also quite readily communicate form, detail, scale, etc. It may also facilitate evaluation and at the same time provoke further generation of ideas. In reality, drawing styles and purposes merge gradually into one another and reflect the personal preferences of the designers, engineers, sculptors and architects themselves. Alan Williams again:

> 'Perhaps in certain circumstances a quality of sketching or scrawling is an indication of a poor or illogical process of thinking but it can reveal a way of using a pencil as a tool to uncover ideas. Few people can actually sit down and draw something they have imagined. It is a natural way of developing ideas. One can usually identify by looking at somebody's scrawlings how hard it is for them to get any ideas. If there is a flow of ideas the sketches, the drawings seem to indicate the lucidity of thinking.'[13]

Research into drawing as a modelling tool owes much to Bruce Archer's work. He stated then that:

> 'Drawing is a very economical way of modelling, it is the fastest and best way of having a quick idea – a visualization – of what is in your head and thus leads naturally into solid modelling.'[14]

The notion of drawing as a modelling device is clearly not new but perhaps the appreciation of its function within the personal and developmental phases of design work is not greatly developed. Modelling through drawing can be a very powerful communicative device, but modelling can have private functions too. It can operate in conjunction with other capacities of the mind so as to evaluate, develop and externalize thought processes of an individual. It is plausible that learning to draw may in some way assist the development of conceptual modelling capabilities – revealing and solidifying for the drawer the nature of form and implications of change. If this is so then surely drawing must be reinstated as one of the major components of design education.

■ References

(1)   Garner, S.W. (1990) Drawing and designing: the case for reappraisal. *Journal of Art and Design Education*, 9(1), 39–55.

(2)   Garner, S.W. (1988) The language of design: drawing on a profound resource. *Studies in Design Education, Craft and Technology*, **20**(3), 133–6.

(3)   Bruner, J.S. (1962) *On Knowing*. London: Harvard University Press.

(4)   Barnes, D. (1976) *From Communication to Curriculum*. London: Penguin Books.

(5)   Powell, D. Product designer, London, interviewed 25 November 1987.

(6)   Webber, C. Painter, Loughborough, interviewed 12 October 1987.

(7)   Axe, R. Director of Concept Engineering, Austin Rover, interviewed 2 December 1987.

(8)   Ballantine, I. Architect and lecturer, Glasgow School of Art, interviewed 20 November 1987.

(9)   Tovey, M. (1986) Thinking styles and modelling systems. *Design Studies*, 7(1).

(10) Williams, A. Director of DCA Design Consultants, interviewed 6 August 1987.

(11) Ashen, P. Then Head of Furniture Design, Birmingham Polytechnic, interviewed 1 March 1988.

(12) McNally, N. Product designer and lecturer, Glasgow School of Art, interviewed 20 November 1987.

(13) Williams, A. *ibid*.

(14) Archer, B. (1976) *The Three Rs*. Lecture delivered at Manchester Regional Centre for Science and Technology, 7 May 1976. [See Article 1.4 in Reader 1.]

4.5

Girls, Boys and Technology

C.A. Brown

■ Introduction

> The class of six-year-olds had some time to spare before lunch and the teacher let the
> children choose what to do. One child asked if they could play with the new
> construction set which was open on the carpet and the teacher agreed. Several children
> at once moved eagerly towards the set. Most of them were boys but one was a girl.
> After a minute or two they were all spread out over the carpet each with enough pieces
> from the set to begin. All, that is except the little girl; she hadn't even managed to
> get near the set and after several tries to break into the circle she gave up and moved
> away toward the bookshelf.

This scene is so familiar that it often goes unnoticed in infant classrooms.
Indeed, even when it is noticed it may be misinterpreted as unimportant unless
some of its implications are considered. These hinge on the connections which
are thought to exist between such play and later achievement in physical
science, technology and related mathematics.

Girls' performance in these areas at secondary school continues to give
cause for concern. There is a large body of research now that documents girls'
rejection of physics at secondary level.[1-5] Their preference for biological rather
than physical sciences is well known.[6,7] This polarization has been a feature of
secondary science education and has been facilitated by the options system
prevalent in the past two decades. It has had the unfortunate effect of allowing
generations of girls to leave school lacking the confidence and qualifications to
pursue job opportunities over a wide section of the employment spectrum. The
hope that the coming of balanced science for all in secondary education will put
this right seems highly optimistic, as the different choices made by girls and
boys are the result rather than the cause of the problem.

Now that the new National Curriculum makes it a requirement that
technology be a part of all children's education it is likely that the problems we
already have with girls' antipathy to physical science will extend to technology. If
this is so we need to identify the causes of these problems and to try to consider
what can be done to alleviate or eradicate them as a matter of some urgency.

■ The importance of leisure experiences

The differences in girls' and boys' leisure experiences have been offered as one of the causes of the polarization of interests described above. It has been claimed that boys' leisure activities are of a kind 'offering greater opportunity to . . . acquire an appropriate grounding for later learning in physics'.[5] These leisure experiences are not simply those which occur when girls and boys are of secondary school age. One can reflect upon the presents given to children of primary school age. The advertising of toys for primary school children strongly reinforces sexual stereotyping and 'girls' toys' and 'boys' toys' are often grouped separately in stores and toy shops. One girl in a class of seven- and eight-year-olds is quoted as saying she preferred Meccano to drawing but 'other people think it's for boys'.[8]

An APU study of the activities which 11-year-old children recollected engaging in over the whole of their childhood indicated marked differences in girls' and boys' prior experiences.[5]

Table 1   Recollection rates for technically related activities.

	Girls	Boys
(1) Making models from kits	6%	42%
(2) Play with electric toys	16%	45%
(3) Creating models with Lego	23%	50%

The APU tests of children's abilities at 11 years showed only one large test score difference between boys and girls: in 'applying science concepts'. The discrepancy was especially marked in those questions dealing with the application of physical concepts; it has also been found in the test results for 13- and 15-year-old pupils and has been found to increase rather than diminish with increase in age. This study by the APU[5] also found that children as young as six years had formed strong opinions about the kind of activities boys may undertake and those which are 'suitable' for girls. Other evidence confirms that these attitudes are not ephemeral and that by the time girls are 15 they have become linked to job aspirations and life choices in a very limiting way.[9]

■ Attempts to alleviate the problems of girls' underachievement in science and technology

Such attempts as have been documented at the secondary stage have been expensive to mount and so far have achieved limited success. However, there have been recommendations in reports of these attempts for action to be taken in the primary school.[10–12] At first sight this may seem easy to arrange given the more flexible conditions of timetable and organization in a primary context. But here too things may not be as simple as they seem. Perhaps some information from the closely related field of information technology might throw some light on the ways in which girls' attempts to gain experience can be thwarted in primary schools.

The MESU project working with third year juniors found that 'girls did want to do more technology but feared that the boys were better at it and felt overwhelmed by them'. The study found that many teachers tended to think that because they offer a choice of activities to all pupils they were giving them equal opportunities. Consequently when girls said they did not want to build models, the teachers concluded that the girls were not interested. But the study team found that the infant girls were reluctant to make Lego models because they felt they would not make them as well as the boys or thought they should be doing 'proper work like drawing and writing'.[8]

After discussion with the researchers about these wishes and feelings 'it was as though some submerged part of them had surfaced . . . . Their teacher was astonished'. The solutions offered by these girls were very revealing. One little girl wanted 'a wishing stone so I could play at the computer and the boys would not be able to choose'. Another wanted 'to have a go early in the morning before register' but added that that would not work as 'then the boys will moan and sulk all day . . . then nobody might be able to have a go'.

Discussion with the children about this inequality of opportunity was followed up by the introduction of a system of working in mixed pairs. So far, this seemed to work well, although a lot of support was required in order to overcome initial shyness and embarrassment, even though the children concerned were used to working in mixed classes. The long-term effects cannot yet be ascertained, nor can it be said that working in mixed pairs in this way has been shown to be compatible with most teachers' preferred teaching styles for this age group.[8]

Studies carried out so far have therefore collected some valuable indicators as to how girls' limited horizons can be established early in life and have begun to describe their performance in some areas and the constraints which can limit it. Information about what girls can do in the pre-secondary stage is still very sparse and a fuller description is required.

But technology in the primary curriculum is virtually a new subject. As recently as summer 1988 Toft said that there was a need for 'a sound body of knowledge about the nature teaching and development of CDT'.[13] In contrast to subjects such as mathematics there is little or no information available from research about what constitutes normal technological development. Such information needs to be based on direct observations of how a specific group of children develop over an extended period of time. Intervention to alleviate gender discrepancies in achievement requires detailed information. First there is a need for longitudinal observational studies to determine the normative pattern of the development of technological thinking in young children. Thereafter the characteristics of gender differences and their causes can be investigated.

■ Two studies of young children's constructional activities

The studies carried out so far offer descriptions of what was achieved by children of two vertically grouped classes containing equal numbers of boys and girls with ages from rising 5 to 6+ years and from 6+ to 8+ years. Observations were made of the children's use of a selection of construction materials. In order to begin to establish a normative pattern of technological development a strategy was put into operation to improve equality of access to the relevant materials.

All the children were taught according to a programme giving them regular structured experiences of the construction sets to be used. This was a modification of the 'free choice' system of access to construction materials where they are seen principally as toys for the children to use in undirected play. If a completely free choice had been given though, evidence of technological ability would have been affected by the extent of different pupils' exposure to the materials. In an extreme case a pupil regularly opting to read or draw at this time might have had no engagement with them at all. So it was decided that specification of a basic amount of 'structured' experience was necessary. The teacher then organized the class so that during each week the mixed sex groups of four to six children all did construction work and she arranged for every child in the class to have the opportunity to gain basic experience of each of the different construction sets used.

The children's work was monitored over a whole school year in each of the two studies and was analysed for evidence of technological development. This was done for the whole class and then the data for girls and boys was analysed separately so that any differences in achievement could be detected.

☐ Study 1

This study of a class of six- to eight-year-old children has been fully reported elsewhere.[14] The construction materials used were a large teacher's set of Lego 1053, a teacher's set of Lego technic and a small Lego switchboard. One one day in each week a BBC micro was available with an In Control interface to enable it to run Lego motors, lights and buzzers. All the models which were completed were photographed together with their makers. In all six sets of photographs were taken recording models made throughout the whole of the year.

Of the total number of models made during the year, many more were made by boys than by girls. At first sight, in a class with equal numbers of both sexes, this seemed to confirm previous expectations and indicate that the girls had been inhibited from making models throughout the year. On further examination however the results were not so clear cut. It was found that the girls had greatly increased their rate of successful completion in the latter part of the year, by which time the number of girls experiencing success in their model making was only slightly less than the number of boys.

In addition to recording the number of models made and the gender of the model makers, the quality of the models made was assessed. This was done by referring to a list of criteria drawn up by the author from descriptions of the structural, mechanical and electrical principles the materials could be used to teach (as described in the Teachers' Guides to the construction materials used).

When the coverage of the criteria was assessed for the whole year it was found that the boys as a group had incorporated far more of the technological features described and had used them with greater frequency in their models. Comparison of the quality of models made at the beginning and end of the year, however, showed that much progress had been made by the girls. For example, by the end of the year many more of the girls' models had features such as wheel sets made up from separate parts, gears and chain drives.

Overall, the results showed that despite the majority of the construction materials used being unfamiliar to the girls at the start of the year, given equal access and encouragement, in a normal classroom situation without any positive

Table 2   Criteria for assessment of technological achievement using Lego 1053 and Lego Technic.

Does the photograph indicate that the maker/s could:

(1) Select and match pieces to make up parts?
(2) Assemble parts to make a stable structure?
(3) Incorporate levers? (e.g. in tip-up trucks or scissors model).
(4) Incorporate wheels to make the model mobile?
(5) Incorporate pulleys?
(6) Use gears to transmit movement?
(7) Use a belt drive to transmit movement?
(8) Use a chain drive to transmit movement?
(9) Use a drive shaft to transmit movement?
(10) Change the speed of movement? (using gear ratios, worm gears, belts or chains).
(11) Control the speed of movement? (using braking system, ratchet and pawl, flywheel or governor).
(12) Change the type of movement? (e.g. from rotary to up and down).
(13) Change the direction of movement? (through right angle for e.g. using bevel gears, crown wheel and pinion, crossed belts or steering system).
(14) Use electric motors?
(15) Use simple electrical switching?
(16) Use switchboard system?

discrimination in their favour the girls became confident enough to take part in such modelling and built up experience of putting technological ideas into practice. After 12 months the performance gap between boys and girls was still there but it had narrowed considerably.

This study therefore raised interesting questions about whether a similar gender gap existed at an earlier stage and if so whether it would be similarly susceptible to the structured access programme which the teacher had been able to incorporate into her everyday teaching arrangement without too much difficulty.

It was decided that this should be the subject of a further study to run over a full school year, starting with the youngest children as they entered compulsory education.

☐ **Study 2**

The class took in children who were either five years old or rising five in September 1987. They were studied throughout the whole of their first school year.

The construction materials used were chosen to suit the age of the children involved. They included a large wooden set of shapes, a set of Fischerform materials which mainly made vehicles, the class tray of odds and ends of Lego (mainly standard red bricks) and a teacher's box of Lego 1053, plus a small set of Lego battery holders and Lego light bricks which were added later in the year. The materials were introduced through a structured access system similar to that described in the previous study. As before, the teacher made sure that all the children in turn were regularly given the time and opportunity to make models with them. The finished models were then photographed with their makers who were asked to demonstrate them and then to display them for a short time. When the colour slides of the models were returned they were also asked to say what their model was and to talk about it; notes were then kept of this information.

The most difficult part of this study came with the need to draw up a list of criteria against which the quality of these models could be assessed. There was no help this time as in the previous study from the Teachers' Guides to the sets of equipment. For this age range a guide often simply takes the form of a few pictures of possible models. None spell out the ideas the materials can introduce as Lego Technic Teachers' Guides do for older children. It was decided that this would have to be done principally by examining the models made, looking for features they had in common which seemed to indicate progress in technological development.

When the photographs of models made by the class were examined it could be seen that some children, faced with a set of construction materials, began by selecting and putting pieces together in two-dimensional arrangements. Other children did not do this, perhaps because they had previous experience of similar materials. These children were able straight away to make a solid three-dimensional structure such as a tower. A few children were able to make models which had a defined internal space such as a walled playground or a house. Some had subdivided the space to sectionalize their constructions, putting internal walls into their houses, arches in their bridges or making their garages multi-storey. In later models, children purposefully incorporated moving parts using 'ready mades' such as the grabs and winches provided in the Lego set. A few had even made up moving parts from basic pieces, making a Big Wheel from the Lego set, or using the girders and screws of the Fischerform to make an extendable arm for a grass verge cutter.

As time had passed, therefore the nature of some of the models changed. A number of complexes were constructed linking several structures together; farms and zoos, factories with buildings, yards and stores. Vehicles were made mobile by using ready-made parts such as complete wheel sets or by self-assembly of axles, wheels and tyres. In the early part of the year such constructions were fewer and smaller. Later on they became larger and more elaborate with several complexes incorporating vehicles constructed for the purpose; delivery vehicles to move in and out of their factory or cars for their multi-storey garage. A few children were able to use the Lego battery and light bricks. They made lights to go in their buildings and traffic lights for a street.

After examination and comparison of all the models made in that year it became possible to compile descriptions of the different features which indicated that their maker had been able to put a particular idea into practice. The descriptions were then used to construct a set of criteria which could be used to assess each model. These criteria, like the ones constructed for the older children in the previous study, were neither hierarchical nor exclusive. Indeed some models were so complicated that they met four criteria while others met only one or two.

Table 3   Criteria for evaluation of models made using construction sets.

---

(1) Model consists of a two-dimensional structure (a flat pattern on a baseboard one brick high or a wall-like vertical structure).

(2) Model consists of a three-dimensional solid structure.

(3) Model consists of a three-dimensional structure with an inside and outside (a hollow structure or an enclosure).

(4) Model consists of a three-dimensional structure which is subdivided into sections either horizontally or vertically.

(5) Model consists of several three-dimensional structures assembled into a complex.

(6) Model is a three-dimensional structure which incorporates ready-made moving parts.

(7) Model consists of a three-dimensional structure which has moving parts assembled from basic components.

(8) Model is powered by mechanical or electrical energy.

Note: A three-dimensional structure is considered to be one in which elevations consist of more than a single part or parts, i.e. a wall of a 'house' must be more than one brick high.

■ Results

□ Trends discerned in the children's model making over the year

The records fell into three sets roughly corresponding to the three terms of the school year. When the children began in their first term, three of the models made were only able to meet the first criterion and had a simple two-dimensional structure. Sixteen of the models met the second criterion and had a solid three-dimensional structure. Only a few had a structure which met any of the other criteria.

By the time the second set was completed in term 2 some changes could be detected in the children's performance. Firstly there were more models completed, 42 in all, 17 more than in the previous set. Four models were simply two-dimensional but there were now 25 which were of a solid three-dimensional construction. A big rise in the number which included an internal space was seen in this term and a number of other criteria showed a rise too but the Lego light bricks and battery holders were only exploited by a few children to power their models.

During the third term of the year the children completed 52 models. The level of production of 2D and 3D structures was similar to the previous term but the proportion of models which included internal space continued to increase and the building of complexes overtook the construction of subdivided structures in popularity. More moving parts were added, both ready-made and self-assembled but the number incorporating the Lego light units and battery stayed the same.

One important result of the structured programme therefore, was the sheer overall rise in the number of models made over the year. Progress was also made in incorporating more of the features requiring application of the technological principles listed in the criteria, the shift from the making of solid models to ones with defined internal space being particularly marked. Contradicting earlier expectations that the children would become bored with the limited range of materials, or that they would repeat themselves, making the same models again and again, it was evident that their work had greater variety at the end of the year than at the beginning.

□ A comparison of the achievement of girls and boys after their first year of schooling

In spite of the practice of assigning mixed-sex groups to constructional activities, there was only one model made during the year by a mixed partnership. A decision was made to leave this one model out of the following analysis as it was not felt to be sufficiently significant to warrant a grouping of its own. When, with the exception of that one model, the total numbers of models made in each term were considered, it could be seen that in spite of there being equal numbers of both sexes the boys made many more models in each term (see Table 4). As a group they seemed to have a confidence and breadth of experience from the start which enabled them to make a wide variety of models earlier than the girls. Every criterion for which suitable materials were available had been met in at least one of the boys' models by the end of the first term. (Note: Lego batteries and light bricks were not available until term 2.)

In the second term the boys more than doubled the number of models they made with a marked rise in the number having defined internal space and in its subdivision. When the Lego battery holders and light bricks arrived at the start of term 2, a small number of boys were also able immediately to incorporate them into their models and it was mainly the boys who continued to use them in term 3.

The building of large complexes was mainly a male-dominated co-operative activity often involving a group of as many as six or seven boys at once. It began in the second term and came to be a regular feature of the classroom in the third term. The complexes included a variety of 3D solid structures, structures with defined internal space and subdivided structures. These complexes were frequently made from several different construction sets at once and gave opportunities for the boys to incorporate movement as some of the complexes required mobile structures such as cars, trucks or trains to make them complete. The use of ready-made and self-assembled parts such as axle/wheel sets also rose rapidly when the boys made a variety of other models unrelated to the complexes, and there was a marked increase in their manufacture of self-assembled moving parts such as lifting ladders.

In contrast with the boys, the girls mainly made simple 2D and 3D

Table 4.

Models made: Terms	By girls 1:2:3	By boys 1:2:3	By the whole class 1:2:3
Number of models made	8:4:16	17:38:36	25:42:52
Criteria met by models			
(1) Two-dimensional	2:2:3	1:2:2	3:4:5
(2) Three-dimensional	6:1:5	10:24:16	16:25:21
(3) Hollow	0:1:5	2:13:15	2:14:20
(4) Subdivided	0:0:2	3:9:3	3:9:5
(5) Complex	0:0:1	4:3:6	4:3:7
(6) Ready-made moving parts	1:0:7	11:20:16	12:20:23
(7) Self-assembled moving parts	2:0:1	1:5:7	3:5:8
(8) Powered	N/A:1:1	N/A:3:3	N/A:4:4

models for the first two terms. Except for two models that were embellished with moving parts which they assembled from basic components, there was little evident achievement of any of the other criteria during that period. It was only in the third term that there began to be a flowering of activity and interest among the girls' group in general. This led to the production of more models and of a range of models which between them met all the listed criteria for the first time.

In spite of this new breadth of achievement it was still true that the girls' performance was quite a long way behind that of the boys, even in the last term of this year. It was not as good by then in terms of the number of models made as the boys' achievement in the first term. Nevertheless the girls' work in the third term showed a variety of achievement across all of the criteria, indicating that they were on their way to acquiring some of the range and depth of experience by the end of the year which the boys' group possessed in their first term.

■ Implications of the studies

☐ For general progress in the teaching of technology in the early years of schooling

The attempt to demonstrate the opportunities offered for technological learning by construction materials in the two studies would seem to have been successful in that it has been possible to devise criteria that distinguish between the models children make. These may aid teachers in promoting children's learning through more purposeful and directed use of these materials. It would seem to be important that these descriptions be confirmed or modified by application to children across the whole early years age range.

☐ For the progress of girls through the teaching of technology in the early years of schooling

The performance of girls in response to a programme of structured access (giving them as near as possible equal opportunity to use model-making materials) was seen to improve both the number and variety of the models they made. It was clear from the analysis of the models that there was a large gap in achievement between the boys and girls on entry to formal education. By the end of a year during which structured access to construction materials had been made available to the rising five- to six-year-olds, the gap seems to have narrowed somewhat, although the girls were still only meeting the criteria achieved by the boys at the beginning of the year. In a six- to eight-year-olds' class the gap between the girls and boys at the end of a year was also narrowed after following this practice. However, it would seem that something in addition to general equality of access is required to enable girls in general to match boys' achievements in this area. One possibility that has emerged from these studies is the introduction of information derived from the criteria which helps teachers to provide activities that promote the learning of technology. It is suggested that this might help to make good the backlog of experience which girls seem to lack and so is worthy of further investigation.

Acknowledgement

With acknowledgement to Ms A. Parker and staff of West Earlham First School, Scarnell Rd, Norwich for their considerable help with these investigations.

■ References

(1)   Department of Education and Science (1990) *Girls and Science*, HMI Matters for Discussion 13. London: HMSO.
(2)   Kelly, A. (1981) Choosing or channelling? In A. Kelly (ed.) *The Missing Half. Girls and Science Education*. Manchester: Manchester University Press.
(3)   The Royal Society and Institute of Physics (1982) *Girls and Physics*. London: The Royal Society/Institute of Physics.
(4)   Kahle, J.B. (ed.) (1985) *Women in Science*. London: Falmer Press.
(5)   Johnson, S. and Murphy, P. (1986) *Girls and Physics*, Reflections on Assessment of Performance Unit Findings (Assessment of Performance Unit Occasional Paper No. 4). London: DES and Assessment of Performance Unit.
(6)   Department of Education and Science (1985) *Science 5–16: A Statement of Policy*. London: HMSO.
(7)   The Royal Society (1985) *The Public Understanding of Science*. London: The Royal Society.
(8)   Milner, A. reported in H. Wilce (1988) Mixed doubles. *Times Educational Supplement*, 24 June.
(9)   Ryrie, A.C., Furst, A. and Lauder, M. (1979) *Choice and Chance*. Edinburgh: The Scottish Council for Research in Education.
(10) Tall, G. (1985) Changes as indicated by examination entries in England. *School Science Review*. 66(237), 668–81.
(11) Johnson, S. and Bell, J.F. (1987) Gender differences in science option choices. *School Science Review*. 69(247), 268–76.
(12) Doherty, M. (1987) Science education for girls: a case study. *School Science Review*. 69(246), 28–33.
(13) Toft, P. (1988) Editorial in *Studies in Design Education Craft and Technology*, 20, 2.
(14) Brown, C.A. (1989) Girls, boys and technology, getting to the roots of the problem: a study of differential achievement in the early years. *School Science Review*, 71(255), 138–42.

PART 5

Implementing the Teaching and Learning of Technology

5.1

The Nuffield Approach to the Role of Tasks in Teaching Design and Technology

D. Barlex

■ The nature of the design and technology task

Consider a design technologist, who observes a situation and perceives needs and opportunities. Exactly what constitutes a need or opportunity is governed by the values of this observer. What is regarded as normal and unavoidable by some will be seen as cause for concern and ameliorating action by others.

Identifying the need or opportunity is only a first step, however, because the design technologist then needs to perceive the possibility for action within the constraints of available resources. Once the need or opportunity is seen as something that can be tackled with the possibility of success then the resolution of the change required can be seen as a TASK. There is of course a wide variety possible in the nature of such a task. It can range from being almost totally concerned with design (as in the designing of a coloured pattern) to the almost totally technological (as in developing the manufacturing system to reproduce that pattern in large carpets).

The design technologist will have responsibility for the nature and quality of the response and will usually tackle the task as a sequence of sub-tasks calling on a range of resources, such as:

- knowledge
- skill
- facilities
- materials and components
- organizations
- people
- time.

There will not be a single uniquely correct solution to be achieved through tackling the task, but many possible responses of varying appropriateness and success. Different design technologists may well come up with different but equally appropriate solutions.

In the 'outside school' world there is a whole range of professionals with pertinent knowledge, skill and values poised ready to swing into action in tackling the task. This is not the case in school. Pupils need to be taught the knowledge, skills and values as well as being put in the position of being a design technologist. This will reveal their capability in 'operating effectively and creatively in the made world' and showing 'increased competence in the indeterminate zone of practice'.[1] It will be by manipulating the tasks that pupils tackle that teachers will be able to exert influence over pupils' design and technology education and provide for progression and differentiation.

In setting design and technology tasks it will be important to be clear on the educational aims of the task and possible outcomes. The totally open situation in which pupils are wholly responsible for finding and tackling a wide variety of tasks is to be avoided for several reasons:

(1)   It is very difficult to resource the pupils' work in terms of materials and equipment as many of them may have different requirements.

(2)   It is very difficult to meet the teaching requirements of each pupil as every task may be very different. The teacher is reduced to 'fire-fighting' on an individual and *ad hoc* basis.

(3)   It is almost impossible to have any clear overall learning intentions for the work matched to the Programmes of Study (or syllabus).

(4)   The pupils are likely to be confused about the requirements of the task and fail to meet important features of both the Programmes of Study (syllabus) and the Attainment Targets (or other assessment criteria).

A further factor which can upset the educational aims of the task is the possibility of different perceptions of the task by the pupil and the teacher. This can arise from the pupil having such a different view of the underlying need or opportunity that what constitutes a solution is open to debate. Also, pupils' perceptions of the task will radically affect what resources they choose to bring to bear on the task. Unless the teacher is sensitive to this there is always the possibility that any interventions by the teacher will be misguided and counterproductive.

■ Two types of tasks for design and technology teaching purposes

The Nuffield Design and Technology Project has identified two broad categories of tasks that can be used to meet its teaching purposes. The first category is *resource tasks*, which are designed to help pupils acquire the knowledge, skills and values necessary for capability in design and technology. There are many different types of resource tasks but the key feature of them all is that they have clear and definite teaching intentions. The following selection of types will give you some idea of the variety possible:

- vocabulary – these require pupils to explore and appreciate existing solutions to standard problems. Through these, pupils become familiar with processes, techniques, materials, components and devices. These features become part of *their* design and technology vocabulary.

- step design – these require pupils to tackle detailed, tightly controlled design tasks using a small part of a single area of technical knowledge. Through these, pupils can build up confidence as they are not confused by a wide range of multiple factors.

- design scenario – a small piece of fiction describes a human situation which needs a design and technological response. Through these, pupils can gain further confidence by using their understanding from one or more areas of technical knowledge.

- role play – putting pupils into small discussion groups with clear guidelines as to their 'job' and value position. Through these, pupils can learn to listen to and appreciate the views of others.

- games and simulations – engaging pupils with decision-making in situations normally outside their sphere of operation. Through these, pupils can develop an understanding of strategies and tactics of design and technology.

The second category is *capability tasks*, and there are two types:

- identified open task – here pupils tackle a complete design and technology task, which has been identified within a task setting created by the teacher, developing a response to the point where it can be put to the test. Through these tasks, pupils have the opportunity to develop and demonstrate capability in design and technology. Note that the task is 'set' by the teacher so that even though it is a capability task it can have clear learning intentions.

- spontaneous open task – here the pupil again tackles a complete design and technology task but the task has been identified completely by the pupil. Through these tasks, pupils have the opportunity to demonstrate the highest level of capability.

Both capability and resource tasks are necessary. Capability tasks without the underlying foundation provided by resource tasks will reduce pupils to operating on little more than heightened common sense. Resource tasks do not directly teach pupils to be capable. Good design and technology teaching will be based on the interplay of these two sorts of task. It will be important to let pupils know the purpose of the tasks they are tackling.

■ The demands that can be made by the different features of tasks

These will be considered under four headings: technical knowledge and understanding; making skills; procedural skills; and range of issues addressed.

☐ Technical knowledge and understanding

If pupils are to gain technical knowledge and understanding through resource tasks and then use (and perhaps extend) this in capability tasks, it is important that teachers take into account the concepts required for such understanding and the level of demand placed on these concepts. A simple model for pupil development within an area of technical understanding requires progress:

- from qualitative appreciation, acquired through simple vocabulary tasks;
- through qualitative application, acquired through simple step design and design scenario tasks;
- to quantitative appreciation, acquired through more demanding vocabulary tasks;
- to quantitative application acquired through more demanding step design and design scenario tasks.

The Nuffield Design and Technology Project sees the provision of sequences of resource tasks to enable such progress to be made in a range of areas of technical understanding as an important strategy in providing pupils with a growing bank of resources that enables them to produce technically more sophisticated solutions to design problems. It is clearly important that pupils are made aware of the purpose of any resource tasks they are required to tackle in terms of their likely usefulness with regard to a current or imminent capability task. As pupils become more competent, self-directed learners, they can be encouraged to choose and even devise their own resource tasks.

☐ Making skills

If pupils are to make what they design it is important that they relate their designs to both their level of making skill and the range of making possibilities at their disposal. This requires the teacher to be clear on when to introduce making skills to support designing and making. It is possible to pre-empt the need for some particularly difficult making skills by providing ready or partly prepared materials. The use of some machine tools or equipment can also reduce tedium and drudgery. There is no substitute for clear skill instruction and the opportunity for practice. The Nuffield Design and Technology Project sees this as forming an integral part of pupils' experience. It is important to make appropriate demands on accuracy and precision in the first instance of using new skills. Guidance in designing is of particular importance, as with suitable intervention a good but difficult (if not impossible) to make design can be adjusted through negotiation to one that requires the pupil to exercise skill but not at such a high level that failure is almost inevitable. Or, worse still, that the teacher has to take over, and carry out the operation for the pupil.

☐ **Procedural skills**

These form the strategies and tactics of design and technology activity and the Nuffield Design and Technology Project believes that pupils can and should be taught such skills quite deliberately as resource task exercises to improve individual competences as well as part of capability tasks. The following list shows the range of strategies and tactics with which pupils should become familiar:

- ways of exploring problems
- ways of collecting data
- ways of modelling
- ways of planning
- ways of evaluating
- ways of working with other people.

Clearly such skills can be used and taught at different levels. Capability tasks can be put in settings that make greater demands on some of these procedures than on others, so that pupils extend themselves by choosing to use those procedures most appropriate for developing responses to each task.

☐ **Range of issues addressed**

One of the most important features of a capability task is that it is balanced in the sense that it draws on knowledge, skills and values appropriately throughout the task. This can to some extent be achieved by means of the task and its setting but it is still possible for pupils to skew their response. A pupil can become so technically engrossed in tackling the task that the inappropriateness of the solution as far as the potential user is concerned is not apparent. A pupil can become so wrapped up with the knowledge of the problem and the subtleties of the human need the solution should address that a made outcome never emerges. A pupil may become so over-concerned with making, that the technical working and purpose of the item being made becomes obscured. Only by helping pupils to inform the use of knowledge and skills with value judgements will teachers be able to ensure that an appropriate range of issues is addressed throughout the task.

■ Task presentation

The way a capability task is presented to pupils will have a significant effect on their perception of it and whether they see it as worth tackling. In the Nuffield Design and Technology Project a large emphasis is laid on both the setting for the task and the stimulus that is used to introduce it. The setting is a description which puts the task in a definite situation. This situation should be described sufficiently well for it to have a ring of truth and air of reality. It should also enable the pupils to begin to identify those features of the problem which are important and those which are trivial or insignificant.

It is also important that the task plus the setting make it quite clear to pupils what is expected of them in terms of final outcome. Only if the required nature of the final outcome is clear will pupils be able to plan effective use of their time in tackling the task. Problems arise when there is a mismatch between the pupil and teacher perception. The Nuffield Design and Technology Project makes provision for capability tasks to be stopped or started before manufacture either at the initial or developed design proposal stage. Clearly such shortened tasks will not develop or reveal total capability to the same extent that a complete task does but they may provide the opportunity for developing capability in the early or later stages of designing and making.

The combination of the task, its setting and the main aims of the task inform how it can be presented to pupils. Below is an example of a capability task, with its setting and main aims, taken from the Nuffield Design and Technology Task File, to illustrate how this might be used to present the task in a way that gives pupils both access and ownership.

☐   Task

To use the resources of design and technology to develop a novelty product suitable for sale as a small gift or souvenir to prototype stage.

☐ **Task setting**

Novelties Inc. is a firm that specializes in the production of small gift and souvenir items. Typical of its clients is a garden centre which has a butterfly farm and aquaria. The task is to develop products that could be sold at the centre's gift shop. (Other possible settings are a zoo, museum, stately home, theatre, summer fair, etc.)

☐ **Main aims**

(1) To enable pupils to design simple products with aesthetic appeal.

(2) To enable pupils to produce these products as prototypes to a good standard of finish.

(3) To enable pupils to consider the costing and batch production of such items.

A range of possible and typical outcomes that pupils are likely to produce in tackling the task is also identified so that the teacher can be clear how the aims are likely to be resolved in terms of designing and making.

☐ **Possible outcomes of the Novelties Inc. task**

- construction materials:
  - flat packed novelties that slot together to form insects,
  - simple jewellery (wood, metal, plastic, enamel or ceramic) based on animal or plant form,
  - simple hanging mobiles of animal shapes
- textiles:
  - printed and stuffed fabric tiles which together form a foliage wall or coral reef,
  - insects or fish that can be 'fixed' to the wall or reef,
  - insects or fish that can be worn as jewellery,
  - decorated scarves
- novelty food products such as:
  - fudge creepy-crawlies
  - piranha biscuits
  - chocolate butterflies
  - snail bread
  - worm cakes

- caterpillar cookies
- graphic media:
    - commemorative book marks
    - beginners' guides to fish or insects
    - pop-up gift cards
    - shaped storybooks for the very young

There are a variety of well tried techniques to introduce a task once a setting has been decided. This can often involve putting pupils in the place of someone within a design-based organization who is being asked to respond as a designer–maker. They can be the recipient of a directive from the managing director or a letter from a potential customer to a firm of design and technology consultants. Television programmes, such as Thames TV's Design and Technology series or SCSST's[2] Technology in Context videos can be used as stimulus for a variety of settings. Visits to school by design and technology professionals who either set tasks (in co-operation with the teacher), or act as consultants to tasks under way, can give a task setting a much needed lift. And of course there is the possibility of a visit out of school to the site of an existing design and technological endeavour or to a situation where needs and opportunities might be found.

■ Task support

Once the pupils are engaged upon a capability task it is essential that they are supported throughout. To provide guidance for this support and to ensure that the main teaching aims of the task are met, the Nuffield Design and Technology Project produces possible teaching sequences which describe the teaching of capability tasks in terms of an assembly of the following elements:

☐ **An introductory stimulus**

This has already been discussed under 'Task presentation'.

☐ **Resource tasks**

These have been discussed earlier. They need to be considered as sequences both within a particular capability task and across a series of different capability tasks. They may occur as a set within a single allotted time-span or at different points within the capability task. It is important that pupils begin to see them as enabling activities well worth some effort.

☐ Task clarification

This is an important feature that can easily be overlooked in the hurly-burly of getting going. Pupils should try to develop an initial specification and ask themselves 'How long have I got?', 'What will I need to do?' and 'What might I need to learn?' They should produce a schedule.

   The pupil's perception of the task will colour what is seen as relevant knowledge, understanding and skill. Many pupils find it difficult to keep their technical options open when designing. It is important that task clarification does not unduly close down the range of responses. While the resource task work provided by the teacher may be aimed at giving pupils a particular technical understanding seen initially as relevant to possible solutions, pupils should be encouraged to think outside this. It is of course necessary to balance this against the practical provision of materials and equipment.

☐ Skills instruction

This has already been discussed under demands made by the different features of tasks.

☐ Preliminary design proposal

This is an important feature to develop from the specification. It should not be over-refined – a simple annotated sketch or model is quite adequate. But without this it is impossible to take stock of the situation and move the task forward in anything but the most haphazard way.

☐ Design development

This is the meat of the designing where the imaging and modelling moves the design idea from the preliminary proposal to a clear and detailed view of the item to be produced. The plans for the production may not always be finished in the complete working-drawing sense, and during production there may well be modifications to the design but it should be the intention of the pupil to develop, through a series of models, the information required for making.

☐   **Production**

It is important that sufficient time is made available for this stage. Hurried work is invariably spoiled as skill instruction is ignored in the rush to finish. Allowing enough time also requires that pupils' design ideas are realistic and capable of being produced by them in a reasonable amount of time. Guidance that bears this in mind during the design development phase is crucial.

☐   **Three review sessions (at least)**

These provide the momentum for the task by encouraging reflection at these critical points:

(1)   after the preliminary design proposal to check the proposal against the task clarification;
(2)   after the design development to ensure that the design will do what is required, can be produced and that materials, equipment and components are available;
(3)   after production to check performance against the specification and user reaction.

It is the intention of the project that teachers should adapt the possible teaching sequences to meet their own situations and to use the elements listed above to construct new sequences suitable for the capability tasks that they develop.

Experience of developing materials has indicated that many teachers would like a suggested teaching sequence for a capability task to be presented to them as a lesson-by-lesson guide using the headings listed above. Experience has also shown that the teaching sequences for a task that can be resolved in different materials are likely to need different patterns of lesson distribution within the sequence for each material. So, while the overall time for the task might be the same, the pattern of lessons for pupils using food will be different from those using resistant materials.

■ Planning for breadth and balance through tasks

The National Curriculum for Technology attempts to achieve breadth and balance through design and technology tasks by identifying in the introduction to the Programmes of Study a set of features that must be present in each Key Stage. *Breadth* implies that there is a range in each of the features that must be addressed. For example, in the case of context five have been identified: home, school, community, recreation, and business and industry. Breadth requires that each of the contexts is met by a pupil in a Key Stage. *Balance* requires that there should be fairly equal amounts of time spent on each. Twenty-five weeks spent on the home and two weeks on each of the others within a single year is broad, but it is not balanced.

Clearly one or two capability tasks will not incorporate all the requirements of the National Curriculum for Design and Technology but a sequence of such tasks can be planned to provide this comprehensive coverage. It is only by planning a course across a Key Stage and auditing the plan against the required materials, contexts, and kinds of outcome, that breadth and balance will be achieved. The Nuffield Design and Technology Project will provide a set of capability tasks in the Task File which allow for choice and variation in providing breadth and balance.

■ Planning for progression through tasks

Here we need to distinguish between pupils on the one hand making progress in resource acquisition and on the other hand, gaining in capability as a result of having more resources at their disposal. It is important that a capability task makes appropriate demands on a pupil's resources. This requires careful matching by the teacher at the level of the whole class, which will take into account the position in the Key Stage and hence the previous experience as well as any overall strengths and weaknesses of the class. Then, while all the pupils in the class are tackling a common task, it will be through individual negotiation and intervention that the teacher will be able to help individual pupils adjust their response to the task so that the demands it makes are appropriate for a particular pupil.

It will also be important to identify those areas of knowledge, understanding and skill that are taught elsewhere in the curriculum and are related to knowledge, understanding and skill useful in tackling design and technology tasks. The approach of, say, science to a particular area – e.g. electrical circuits – will not necessarily generate understanding that is applicable in a design and technology task. Developing a wiring diagram and connecting up the circuit is a very different task to building the working circuit on a simple circuit board. However, a different approach by the design and technology teacher, concentrating perhaps on practical application rather than an underlying explanation, may only serve to further confuse the pupil. So it will be important to check with colleagues in other curriculum areas and try to use a consistent approach even though it will be for different purposes.

There is a wide range of knowledge and understanding, making skills and procedural skills required for design and technology. It is possible to identify many areas in each of these categories and consider progression in each of them. Consider the procedural skill of collecting data. We can consider range, appropriateness and sophistication in this. Range refers to the number of different methods which the pupil can be taught. Appropriateness refers to the pupil being taught when to apply one method rather than another. Sophistication refers to the level of expertise with which the method is used.

The Nuffield approach is to develop resource task sequences which enable the pupil to make progress. For a procedural skill this will teach the pupil about a number of methods and enough about each of the methods to enable the pupil to become more sophisticated in its use. The first trials of materials revealed the importance of starting at a suitably low level in such a resource task sequence. The resource task on costing production started with too sophisticated a view of costing and this, coupled with a demand for some arithmetical skills, left pupils floundering and demotivated. Resource tasks may not be the best way of teaching a pupil when to use a particular method. Consider modelling what a design will look like (as opposed to the way it may work). This is an area which can call on a considerable range of techniques. It is quite possible to teach pupils these techniques to a high level of sophistication through resource tasks, but progress in modelling will only be made if the pupils learn to discriminate between the various methods and use them appropriately. It is likely that pupils' progress in modelling will only be revealed by the way they carry it out within a capability task, although they were taught the techniques through a series of resource tasks.

The Nuffield Task File contains sets of resource tasks dealing with the following areas:

- strategies for design and technology
- designing and making with food
- designing and making with textiles
- designing and making with construction materials
- designing and making machines
- designing and making with graphic media and communication
- trading.

Each capability task described in the Task File will indicate clearly which resource task sequences should be used to support it. In addition, teachers will be able to select resource task sequences to support capability tasks that they design.

Now we have come full circle in this article; at the end of the opening section it was noted that pupil perception of a capability task might well influence the knowledge and understanding they choose to bring to bear on the task. The Nuffield Project is committed to develop resource task sequences that enable pupils to make progress in learning for design and technology. It will be important for teachers to develop ways of presenting such sequences that are attractive, and thus open pupils' eyes to the possibilities of use in the current and future capability tasks.[3]

Notes

(1) Taken from *National Curriculum Design and Technology Working Group. Interim Report* (1988). Department of Education and Science and the Welsh Office. London: HMSO.
(2) Standing Conference on School Science and Technology.
(3) The Nuffield Design and Technology Project is producing the following material for Key Stages 3 and 4:

- Capability Task Files
- Resource Task Files
- Students' Resource Books
- Students' Study Guides
- Teachers' Guides

(KS 3 material available from Spring 1994; KS 4 material available from Spring 1995.)

5.2

Stages in Systems Thinking, Modelling and Realization in School Electronics

D.J. Martin

[This article was written for an international audience before the National Curriculum Technology for England and Wales was finalized.]

■ Introduction

The aims of this article are:

(1) To identify some important potential contributions of electronics to technology education.

(2) To describe a model for problem solving involving electronics which allows this potential contribution of electronics to be realized and which develops in complexity with pupils' developing capability and age.

(3) To discuss some of the sources developed and used in the UK which support the kind of work described.

■ Background

The starting point of this [article] is the developing consensus on key aspects of technology education, as it is perceived in the United Kingdom. In particular, the recent interim report of the Design and Technology Working Group (DES/WO, 1988), preparing for the new National Curriculum in [England and Wales], represents a considered statement of good practice in the UK.

The Working Party has suggested that the following 'attainment targets' describe the range of capabilities present in effective design and technology:

(1) explore and investigate contexts for design and technological activity

(2) formulate proposals and choose a design for development

(3) develop the design and plan for the making of an artifact or system

(4) make artifacts and systems

(5) appraise and process outcomes and effects of design and technological activities.

These are not to be regarded as separate discrete activities nor as discrete stages of design. Rather they are identified aspects of the holistic activity of design and technology which interact and interlink with each other.

The most important reason for including significant work involving electronics in technology is that it develops these capabilities in a powerful, effective and motivating way. Its great versatility allows a wide range of contexts to be explored and a wide range of alternative solutions to be researched and modelled. Realization of electronics is relatively rapid and inexpensive, and simple testing (does it work?) is unambiguous and immediate.

The widespread, pervasive nature of electronics in work, leisure, home, communications, medicine, etc. are clear indicators of the power and versatility of electronics. What is less widely appreciated is the simplicity of the key central concepts and skills.

However, the potential benefits of electronics to technology education have been widely neglected. This is partly due to the widely held view that a detailed technical knowledge of electronics is essential before students can engage in any kind of problem solving in its use. This [article] seeks to counter this view and to argue that electronics design is a straightforward, exciting and powerful process which is accessible to a wide ability range if students are encouraged to concentrate on what they plan to achieve at the early stages rather than on the details of how they might achieve it.

■ Stages in electronics design

The following analysis of the stages in electronics design is based on industrial procedures which have subsequently been adopted and adapted in education.

The first stage in using electronics does not consist of investigating and learning about components and circuits. The first stage consists of identifying what is needed and writing a clear specification or brief stating what the system needs to do (not how it will do it).

The next stage is to identify suitable sub-systems which might achieve this specification. The classic systems diagram is shown in Figure 1. In practice

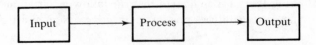

Figure 1 The basic system diagram of any electronic system.

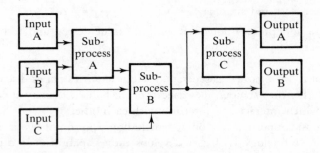

Figure 2 A typical sub-division of the basic system diagram.

each of these units is broken down into sub-units where the actual sub-units used and the signal paths connecting them mirror the specification. Figure 2 shows the kind of sub-divisions typically involved.

The reason why many people imagine electronics to be very complicated is because they do not see this level of the design – they see complicated boards with complicated-looking components. Actually the range of input, output and process sub-systems (described at the level of function) is modest (see Table 1).

Having decided on the best arrangement of sub-systems (and usually there will be several possible alternative arrangements for a moderately complicated specification) the next stage is to decide on suitable components for each separate sub-system. This is where most of the detail involved in the training of graduate electrical engineers occurs but actually most common sub-systems can be adequately 'translated' into quite a modest range of components.

The final level of detail is the physical arrangement of the chosen components, usually as part of a printed circuit board. Then the electronics engineer builds the circuit, tests the sub-units and then tests the system as a whole (Figure 3).

In industry the various stages would typically be the responsibility of different groups: senior management, project management, engineers and technical staff.

In moving from the specification to determine the sub-system needed and their signal interconnection, the key tool is language. The electronics designer looks for clues for the appropriate sub-system in the language of the specification. Of course, at a professional level, sophisticated analysis tools – Boolean algebra, Karnaugh mapping, etc. – are employed but, in fact, language will allow most problems to be tackled successfully.

Table 1 Common input, process and output sub-systems.

Common subsystems		
Input sensors	Process units	Output devices
Movement (switch)	NOT	Lamp
Light	AND	Buzzer
Temperature	OR	Sound
Magnet	Amplify	Movement
Pressure	Compare	(Motor/solenoid)
Rotation	Delay	Radio waves
Proximity	Count	Heater
Radio waves	Remember	Display
Radiation	Pulse	(Numbers/words/pictures)
	Add	

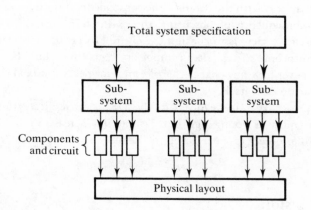

Figure 3 The conceptual levels of electronic system design.

In transferring from the sub-system to the components the electronics engineer characteristically accesses information (textbooks, data books, magazines, etc.). Moving from the components to the layout the main skill deployed is specialized topographical manipulation. In building the systems, the skills employed involve construction and testing.

I want to particularly emphasize the importance of testing each sub-unit as it is built. It is actually relatively easy to test a sub-unit, but it is more difficult to isolate a fault if the entire system has been completely built (Figure 4).

Now, like any model of a human process, this model is an oversimplification. In practice, at each stage the engineer uses knowledge and experience acquired in other projects from other levels of the process. But if this is a

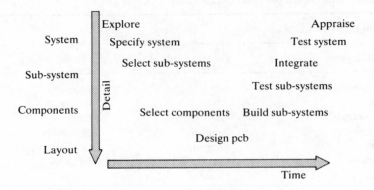

Figure 4   The progress of electronic design.

broadly correct description of the total process of design involving electronic technology, then it is crucial to ask where we should start in schools. The 'traditional' answer is with the details – the components, the circuits, etc. – but this has the consequence of obscuring the essential process.

An alternative view is to consign the entire process until a relatively late stage in young people's lives (after compulsory schooling?) but then the large majority will never have any chance to make an informed vocational choice or to appreciate the process itself.

The approach that has evolved in the UK is to replicate this process, starting from quite a young age but, as pupils increase in maturity and experience, to proceed to finer levels of detail.

■ Progression

It would be dishonest to pretend that a single consistent philosophy exists and guides the establishment of a coherent scheme of progression in the United Kingdom. What I am seeking to describe is an environment in which a range of activities occur and in which there is debate (sometimes heated) but which, in many cases, does offer pupils the kind of deeper levels of experience described (Figure 5).

Table 2 summarizes the range of activities, knowledge and understanding, skills, attitudes and values that, in the author's judgement, represent a coherent pattern of educational development involving electronics appropriate for particular age ranges. Having said that, it is important that individual pupils should be able to go to a depth which is appropriate for their own level of understanding and maturity.

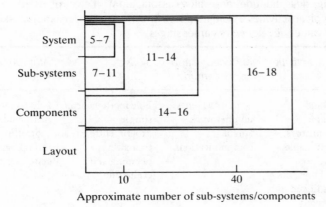

Approximate number of sub-systems/components

Figure 5   Educational progression in electronic design − a possible scheme.

■ Approaches and resources

The following section describes some kind of the resources and approaches currently being used and developed in the United Kingdom. The list is not intended to be exhaustive. Rather it seeks to illustrate the kinds of activities which have been outlined earlier and is chosen because it represents materials with which the writer is familiar.

☐ Age: 5 to 7

At this stage, I would suggest, pupils' involvement with electronic technology should largely be with total systems − electronic toys, programmable toys, remote control toys, etc. One widely used resource in primary schools is floor turtles (perhaps linked to a Concept Keyboard) and later linked to simple Logo commands. The principle aim of this work is the development of number, language and communication capabilities, but there is a benefit in pupils' confidence in their ability to control technological systems.

☐ Age: 7 to 11

In this age range there are two important activities which are widespread in the UK. One is simple experimental 'science' work with batteries, wires, switches, etc. but which can be (though sadly is not usually) linked to modelling activities.

Table 2 A detailed overview of progression in activities, knowledge, understanding, skills, attitudes and values associated with electronics in schools. At each stage, typical activities would reinforce knowledge, understanding, skills, attitudes and values developed at earlier stages.

Age	Typical activities	Knowledge and understanding	Skills	Attitudes and values
5–7	Electronic keyboards, electronic toys, programmable toys, remote control toys, floor turtles linked to Logo, word processing, Concept keyboard	Spatial awareness, number work, language, communication	Fine motor control, simple experimenting, estimating, planning and evaluating	I am comfortable and confident in controlling technological systems
7–11	Simple circuits, batteries, wires, switches, motors, bulbs, buzzers, electromagnets. Electronic sensors and output devices linked to models and artifacts controlled by electronics processing 'blocks' and computers (with pupils making the control decisions)	Science knowledge. A range of output devices can control artifacts using electronic systems	Researching, planning, selecting, constructing, making, fault-finding, evaluating	I am comfortable and confident in handling and connecting electrical and electronic building blocks in order to achieve an outcome I have planned
11–14	As for 7–11 but in more demanding problems using a wider range of input and output devices and processing units and more sophisticated computer commands. 'Translating' a simple system model into a printed circuit	Artifacts and systems can be monitored and controlled using electronic systems because they are able to sense, act, make 'decisions', to generate sequences and 'remember'	Systems thinking. Researching potentially suitable sub-systems. Considering alternative solutions. Etching, drilling, soldering. Checking signal levels to and from sub-systems and power connections to identify faults	Electronic technology is a valuable resource in meeting human needs. I can solve a wide range of problems by applying my knowledge and skills of electronics. I appreciate that electronics is not always the best answer

Table 2   *continued*

Age	Typical activities	Knowledge and understanding	Skills	Attitudes and values
14–16	Tackling problems where research is needed to establish whether or not electronics is a useful resource for the total design. Using computer-aided design to produce printed circuits	Control can be effected by a variety of technologies. Knowledge of the types of situation where each aspect of technology is particularly suitable. Understanding of the importance of feedback in control systems. Understanding of distinction between and merits of digital and analogue systems	Interpreting circuit diagrams for electronic sub-systems. Using probes and monitoring voltages on components to identify faults. Monitoring signal variation with time on an oscilloscope to check sequences	I am capable of solving a wide range of problems by applying my knowledge and skills in technology. I can use electronics in combination with other technologies. I can decide when electronics is and is not an appropriate part of my design

There is also increasing use of Logo, together with various interfaces to control models that pupils have made from construction kits such as LEGO[1] or using an elegant constructional technique developed by David Jinks (Williams and Jinks, 1985). All of these activities are most effective when they are linked with topic work and incorporated within a theme. Experience suggests that it is important that pupils have already encountered 'ordinary' work with Logo and simple model-making prior to using these for control. The main problems are due to the disruption of having to dismantle construction kits on the one hand and the difficulty of building reliable mechanisms (pulleys, gears, etc.) with the David Jinks' method on the other.

A recent UK development is Electronic Sentences, which are a series of small boxes with phrases on them like 'If it is light . . .', 'If a switch is pushed . . .', 'and', 'or', '. . . switch on the bulb', '. . . sound the buzzer', '. . . turn on the motor'. Pupils join these boxes to form simple sentences, e.g. 'If it is light . . . and . . . if a switch is pressed . . . sound buzzer'. The system has obvious benefits in developing language and communication as well as in developing technological capabilities.[2]

☐   **Age: 11 to 14**

In my own judgement, the most appropriate activity in this age range is a broadening of the range of sub-systems with which pupils have experience, together with a limited introduction to simple electronic construction with components. In principle, rather more work with components is desirable but the practical difficulties are considerable.

'Microelectronics for All 3' was the author's first venture into curriculum development and allows pupils to tackle a wide range of 'decisions', 'counting' and 'remembering' types of problem. In a recent total rethink of the support material, all of the problem-solving exercises have been placed in the context of themes: hospitals, safety in the home, animals and pets, etc. Microelectronics for All is in use in approximately 68% of UK secondary schools.

Programmable Systems bears many similarities to the kind of work with LEGO® and Logo described in the 7–11 age group but uses a particular language (Legolines) specifically designed for control work.

☐ **Age: 14 to 16**

A wide range of resources has been developed specifically for this age range in the United Kingdom. System Alpha[3] and the E and L Electronics Systems Kit[4] allocate one small board to each sub-system and so a very wide range of different systems can be synthesized.

In the writer's judgement, the most difficult challenge in this age range is supporting the move from sub-systems to components. As part of a curriculum development exercise, entitled Microtechnology Resources, a 'Modular pcb System' was developed where, for each sub-system used by pupils in their prototype systems kit, there is a corresponding rub-down transfer which can be stuck to copper-clad board, etched, drilled and components can then be soldered in place. This certainly reduced some of the need for complex detailed design. What we failed to adequately stress in the support materials at the time was the importance of building each sub-unit and testing it individually before going on to build the next sub-unit.

I am convinced that the best way forward in this area lies in exploiting the power of computer-aided design. Certainly this is the way in which the industry has gone and it has the potential of removing from students a great deal of the obscure and error-prone detail associated with this kind of work.

■ Conclusions

(1) A clear educational philosophy for electronics and technology involving a movement from systems to sub-systems to components has been developed (albeit in an *ad hoc* manner) and adopted in many UK schools.

(2) The most important 'missing link' is really effective curriculum materials to make the transition from sub-systems to components and circuits as straightforward as is the move from systems to sub-systems.

(3) The most serious need in the UK is teacher training to enable teachers to fully exploit these developments.

Notes

(1) LEGO® Logo and Programmable Systems are available from: Dacta Educational Division, LEGO Systems Inc, 555 Taylor Road, Enfield, Connecticut, 06082 USA.
(2) Electronic Sentences are available from: Educational Concepts Ltd, 16 Heathfield Rise, Rishworth, Sowerby Bridge, West Yorkshire, UK.
(3) Microelectronics for All, System Alpha and Microtechnology Resources are available from: Unilab Inc, 9102–1 Industrial Drive, Manassas Park, VA22111, USA.
(4) E and L Systems Kits are available from: Interplex Electronics Inc, 70 Fullerton Terrace, PO Box 1942, Newhaven, CT 06512, USA.

■ References

Department of Education and Science and the Welsh Office, DES/WO (1988) *National Curriculum Design and Technology Working Group. Interim Report*. London: HMSO.
Williams, P. and Jinks, D. (1985) *Design and Technology 5–12*. Brighton: Falmer Press.

5.3

What is Cooperative Group Work?

H. Cowie and J. Rudduck

■ The distinctive potential of group work

We found little evidence that teachers shared a common conception of the distinctive features of group work as a form of learning. In our view, whatever the form in which group work operates, there are certain fundamental principles that must be respected. The whole point of group work, and its central feature, is *the opportunity to learn through the expression and exploration of diverse ideas and experiences in cooperative company*. It is cooperative in the sense that no-one in any one working group is trying to get the best out of the situation for him or her self; it is not about competing with fellow members of the group and winning, but about using the diverse resources available in a group to deepen understanding, sharpen judgement and extend knowledge.

Groups that are working effectively will have the following characteristics:

- group members are, between them, putting forward more than one point of view in relation to the issue or task that confronts them;
- group members are at least disposed to examine and to be responsive to the different points of view put forward (in role play and simulation, this will occur implicitly during the event but can be made explicit in the review or debriefing);
- the interaction assists with the development of group members' knowledge, understanding and/or judgement of the matter under scrutiny. (See Bridges, 1979, pp. 16–17.)

What matters is that teachers frame the task in ways that support the distinctive potential of learning through group work. The task should also be one that all members of the group think worthwhile to explore together.

Effective group work also depends on a shared understanding among the members of the group of various social and procedural rules. Group activity should reflect such values as reasonableness, orderliness, openness, freedom to take risks with ideas, equality, and respect for persons. Quite often the normal experience of pupils in schooling does not equip them for interactions which support such values. It is important that teachers who are keen to develop group work consider (a) ways of helping pupils to understand the social and procedural rules that underpin group work, and (b) provide some opportunity for sustained experience so that pupils begin to appreciate and feel comfortable with the new way of working.

■ Different forms of cooperative group work

Group work has many guises: in classrooms it mainly takes the following forms.

☐ Discussion

Here a larger group of pupils and their teacher, or a smaller group of pupils without their teacher's constant presence, work to share understandings and ideas. The focus may be the interpretation of something which is ambiguous (a picture, a poem, etc.), the sharing of experiences, the pooling of ideas, or the eliciting of opinions on an issue of common concern. Discussions may lead to enhanced individual understanding, or they may require negotiation in the interests of arriving at a group consensus.

☐ Problem-solving tasks

These usually depend on the discussion of alternatives as a medium for constructive interaction. Often the same task is set simultaneously to a number of small groups of three or four pupils, and there may be a final review of solutions with mutual criticism. Alternatively, an overall problem might be identified as a framework in which groups of pupils work on different aspects of the task and the different contributions are then brought together and reviewed.

☐ Production tasks

These are slightly different from problem-solving tasks in that there is usually a concrete outcome: that is, pupils might be working in teams to produce, say, a film, with one team responsible for the research, one for the technical work, one for the programme sequencing, etc. Or pupils might work in small groups to design and produce, say, a bookstand that even if only partially full holds the books upright! Here, like the problem-solving task, there may be a communal, whole group review of the progress of different bits of the jig-saw, or there may be a comparison of products and even a judgement as to the best.

☐ Simulations

In these, pupils take on the situation or task of a supposed real life group. They might, for instance, become a mining community that gathers round the pit head when a pit disaster has been announced. Or they might, in smaller groups, work to a brief, sometimes against the clock, which puts them into realistic competition with other groups: the situation might be one in which different companies put in competitive bids for a contract to design and build, say, a new children's playground. Within simulations, pupils are often free to contribute from their own strengths or perspectives, although sometimes they may be assigned quite specific roles and the simulation then merges into our next category, role play.

☐ Role play activities

Activities in which each pupil is given a character or perspective within the framework of an event or situation. The role becomes, as it were, a mask and the characters interact according to their interpretation of the given role. Roles are usually assigned to reflect different perspectives on an issue or event: for instance the issue of putting antibiotics into animal feed might be debated by a farmer, a vet, a supermarket owner, a doctor, a low-income parent.

■ Techniques involving group work

'Techniques' should be distinguished from the more fundamental 'forms' of cooperative group work outlined in the previous section. Techniques, of which buzz groups, snowball groups, and cross-over groups are the most common, are merely ways of promoting interaction – of getting ideas flowing – in the interests of collaborative group work. They clearly embody the key characteristics of group work in that they bring small groups together to pool ideas in a framework which positively encourages people to interact and to be creative or to take risks with ideas. However, they are limited and are not, in our view, a substitute for the sustained interaction and depth of understanding that cooperative group work supports.

☐ Buzz groups

To provide an opportunity for greater participation by individuals in a large class event, the teacher may invite pupils to turn to their immediate neighbours and in threes or fours spend a few minutes exchanging views about, for example, things they don't understand; things they disagree with; things that haven't been mentioned, and so on. As Jaques (1984) says: buzz groups enable pupils 'to express difficulties they would have been unwilling to reveal to the whole class without the initial push of being obliged to say something to their neighbours. Taken by itself, the buzz group has little meaning. Yet in the context of a large (i.e. whole class) event it can rekindle all sorts of dying embers.'

☐ Snowball groups

The principle here is that groups of two or three people, with a very tightly defined task that they have to discuss in a very short time, form a partnership with another group and compare ideas. The quartet then joins another quartet to form an eight. Sometimes the culminating task is to arrive at some consensus. This is then reported back, from each of the eights, and an overall position arrived at through further exploration. The advantage of this approach is that it prepares pupils to participate confidently in the final plenary discussion. There are two disadvantages in practice. First, pupils may not always be good at managing the negotiation which allows the pairs to reach a common view and then the quartets to reach a consensus. A consensus may be arrived at through domination rather than through reasoned discussion. Second, pupils may switch off as they find themselves going over the same ground. To avoid this second problem, the sequence of stages can be differentiated. (See Table 1 below which is based on Jaques, 1984, p. 92.)

Table 1 Example task – to find out whether there is a shared perception of school rules.

Stage 1 0 + 0	Brainstorm, and record your ideas in a list	
Stage 2 00 + 00	Share lists and select six that all four of you agree are school rules	
Stage 3 0000 + 0000	Share lists; identify those rules that are common to both lists and list them in order of importance. Appoint a spokesperson	
Stage 4 00000000 + 00000000 + 00000000	Plenary session: review reports and discuss areas of agreement or disagreement	

Here, the task at each stage is slightly different and the danger of repetition is to some extent avoided.

☐ Cross-over groups

If teachers find that the final plenary sessions are not working, or that they need an alternative structure to avoid repetition, the cross-over system can be useful. Pupils are divided into groups on a mathematical basis: for example, if there are 27 pupils in the class, there will be nine groups of threes (three As, three Bs, three Cs, three Ds, three Es, three Fs, three Gs, three Hs, three Is). In order to consolidate the sense of 'home-base', the A, B, C, etc. groups will need to meet for a briefing before mixing so that they establish their identity and control over the agenda. Groups then reform as threes (an A, B, C in each of three groups; an E, F, G in each of three other groups, and so on). Then, after a specific time on their task, they go back to their base of As, Bs, Cs, etc. Again the task needs to be one that can be handled within this format, and also one that gives responsibility to the home-base teams for reviewing the different experiences that the three members will have had in separate groups.

■ Forming groups

Whereas the logic of the lecture is easily translated into familiar and relatively unambiguous roles and physical structure, group work is more diverse. Group size may vary but it should not be so large or so small that it undermines the fundamental principle that group members may learn through easy interchange with each other. Groups that set out to use discussion to explore ideas or experience will not normally work effectively with more than 12 members, while six to eight is often thought of as a good – though in education not always affordable – size. But where interaction supports the pursuit of a problem-solving task or product-oriented task, then an effective group size is usually three or four.

The physical placing of participants should be such as to allow easy interaction: a circle, square, rectangle or arc of chairs will reflect the kind of interaction that discussion is designed to foster. In one classroom we actually saw working groups of pupils sitting in a straight line of four desks – with the result that those at the extreme ends had to bellow in order to communicate with each other!

Groups may be formed from volunteers or may be constructed in order to ensure that different kinds of experience or skills are represented. Teachers are now much taxed by the issue of whether small groups should include both boys and girls and whether they should bring together pupils of different cultural background.

■ Discussion-based group work and the teacher as chairperson

Discussion groups may or may not, depending on pupils' experience of this form of learning and their capacity for disciplined activity, benefit from having a teacher as chairperson. In our experience, teachers underestimate the complexities of discussion-based group work and tend not to appreciate the expertise required of both group members and the group chairperson. Pupils may need to experience good chairing in order to learn how to behave constructively in group work. Although the chairperson is not the fount of wisdom, nor intended to be the main contributor, the role is not an inactive one, for the chairperson will try to see that discussion flows freely across the group and not just between the chairperson and individual members.

The chairperson will encourage non-contributors without bullying them, and indicate that their comments will be listened to and treated with respect. The chairperson will help students to avoid arriving at easy judgements and try to understand views that other group members express. The chairperson will generally favour a thoughtful style of discussion rather than an argumentative style where members are quick to arrive at judgements and are bent on defending entrenched positions. The chairperson will help pupils to express their views freely and without ridicule. The chairperson will be a good listener, will appear supportive but firm, and will signal that all members of the group are important.

The main types of spoken contribution to discussion made by successful chairpersons seem to be these:

- asking questions or posing problems;
- clarifying, or asking a group member to clarify, what has been said;
- summarizing the main trends in discussion;
- keeping the discussion relevant and progressive;
- helping the group to use and build on each other's ideas;
- helping the group to decide on its priorities;
- through careful questioning, encouraging reflective criticism.

Of course, as pupils learn to work effectively together they will assume and share responsibility for these functions. But too often, in our experience, pupils are expected to learn in groups without adequate briefing about procedures, roles and responsibilities.

There is a difference of opinion concerning the need for a chairperson. On the one hand, a chairperson can improve the quality of discussion by acting as a model of an enquirer who shows respect for other people's contributions. But on the other hand, a chairperson who is a teacher and who is perceived by pupils as behaving in an authoritarian or otherwise teacherly manner may inhibit the easy flow of talk – in particular, the kind of talk that is expressed in a register that the participants find they can most easily handle. Barnes and his colleagues therefore explored the use of small, highly interactive groups which gave the pupils more freedom to construct their own meanings and to work things out in their own ways. They concluded, after analysing tape-recorded discussions, that without a teacher present pupil interaction through talk is very different. Pupils are more ready to embark on an enquiry using language to test out the limits of their knowledge and to reshape it collaboratively, often drawing on personal experience as a way of making progress in understanding (see Barnes, 1976).

Aware that teachers who act as chairpersons in discussion often find it difficult to relinquish familiar patterns of classroom behaviour, Stenhouse, in a curriculum project developed in the early 1970s, offered the support of a strong self-monitoring framework. Teachers were encouraged to tape record discussions and listen to them, bearing in mind questions such as these:

- To what extent do you interrupt pupils while they are speaking? Why and to what effect?
- How many silences are interrupted by you?
- Are you consistent and reliable in chairing? Are all the pupils treated with equal respect?
- Do you habitually rephrase and repeat pupils' contributions? If so, what is the effect of this?
- Do you press towards consensus when consensus is inappropriate?
- Is there evidence of pupils looking to you for reward rather than to the task?
- Do you generally ask questions to which you think you know the answer? (Stenhouse, 1970)

Bridges (1979) observes that it is difficult to conduct discussion in school if pupils are afraid to speak freely, if teachers do not think that pupils' opinions are worth attention, if pupils are rudely intolerant of opinions that they disagree with, if pupils feel it improper to express a personal opinion, or if pupils are not amenable to the influence of reason, evidence or argument. Bridges adds, wryly, that in our education system the conjunction of all these conditions is not at all uncommon! We think, however, that the situation is changing.

■ Small group work and the teacher as facilitator

The role of chairperson of a discussion group, as we have seen, requires sensitive handling. So does the teacher's role when the class is broken down into a number of small working groups.

A teacher who has in his or her charge the progress of a number of small groups of children has first to be confident that the tasks set are clear and engaging. Pupils must understand the structure and timing of the session so that they can pace their work – they must also understand what demands will be made of them for sharing their work or reporting back on what they have done.

A common phenomenon is for teachers to intervene either too much or in the wrong way. Galton (1981) (which although based in primary schools is relevant here) documented a dilemma: the teacher, it seems, when 'calling in' on a small group at work did not listen long enough to know where the group was up to, and his or her interventions consequently tended to move the pupils' thinking backwards or to divert it. The research showed that the teacher asked more intellectually challenging questions in whole class activity, presumably because he or she knew where the corporate understanding of the group was at. The lesson for cooperative group work is twofold: first, wait until you are called – without feeling guilty about not doing your job as a teacher – and help pupils to feel comfortable about seeking advice when they need it; or, second, abandon the idea of 'getting round' each group in turn and spend long enough with the groups that you do join to make sure you understand the pattern and direction of their thinking. If the teacher plays a 'drop-in' role, pupils can perceive him or her as fulfilling a policing function – always checking up because of some deep-rooted mistrust of pupils' capacity to get on on their own.

Some pupils in a local comprehensive school, who for the first time had the experience of working collaboratively in small groups on projects in integrated design, made the following comments about the teacher's attempt to manage his new role:

> 'If only he (the teacher) would let us get on with it and not keep calling round.'
> 'If he didn't have so many talks and let us get on with it.'
> 'The staff should stop talking and let us get on with it.'

For their part, pupils who are working in small groups that are not under the constant surveillance of the teacher tend to experience or exhibit the following problems:

- reluctance to value the contribution of some members of the group;
- the persistent dominance of one or two pupils. Sometimes the dominant ones may do most of the talking (boys are sometimes dominant in talking in a mixed gender group) or they may assert their dominance by assigning

low status tasks to other members of the group while maintaining higher status tasks for themselves;

- the breaking up of the group into two slightly hostile sub-groups or two completely separate sub-groups;

- the tendency to proceed from a particular perspective, without adequately examining the alternatives that are available within the group member-ship. This results in one-sidedness or conservatism;

- the failure of one or two pupils to allow themselves to become part of the group. Other pupils in the group sometimes accuse such pupils of being idle, or of being passengers and doing none of the work;

- inequality in the acceptance of responsibility for the group's work or aspects of it. This is sometimes a product of other forces at work in the group (see the first three points above);

- the acceptance of an over-easy agreement in the face of complex issues. This is often because pupils are anxious about the time and concerned to 'finish' before the bell goes, or because they have not been helped to see uncertainty as a legitimate state;

- inactivity and lack of interaction, whether arising from uncertainty, resistance or boredom; boredom may be caused by a task not engaging the pupils or the pupils failing to perceive the depth and challenge of the task as set.

■ Role play and simulation and the teacher as coordinator or participant

'I could never do role play with my class! I'm much too introverted.'

Drama teachers have long realized the potential of role play for helping children to gain an understanding of their own lives, to increase their capacity to explore the hypothetical, to learn about how other people think and feel, and to come to terms with disturbing emotions like anger, fear and hate. Not only does this learning take place within the safety of a playful situation, but also there are other people involved in the process, whether as participants or as onlookers.

Spontaneous role-taking is an activity which appears early on in life. Pre-school children in their play will readily take roles which reflect familiar everyday experiences (mother and baby, doctor and patient, shopkeeper and customer), explore hypothetical situations (robbers and policemen, cowboys and Indians) or enter into a world of fantasy and make-believe (monsters, dragons, witches and princesses).

Drama teachers recognize that we do not 'grow out of' role play but can continue to respond to situations which enable us to take the other person's point of view, explore the outcomes of our actions on others, and experiment with hypothetical situations. All of these are essential characteristics of cooperative group work. With age, however, also comes the ability to stand back and reflect on what we have been doing. Furthermore, since role play engages the emotions, it is often a method which students find motivating and involving.

While in drama lessons it is recognized that role play facilitates both self-expression and self-discovery through work with others, it has not traditionally been seen as an appropriate method in other subjects. We would argue however that it has great value as a teaching method right across the curriculum. In fact, the school's habit of compartmentalizing drama as a subject rather than as a way of learning has meant that many teachers consider that they lack the expertise or the personality to integrate it into their lessons.

As in other forms of cooperative group work, the teacher does not play the part of expert. Rather the students themselves take decisions, since role play with others enables them to build on the experiences, feelings and knowledge which they already have.

There is no one way of facilitating role play and simulations. All would agree that the teacher needs to stand back from the activity in order to observe, reflect and interpret the interactions which are taking place. Some would be content simply to share with their students the perspective which comes from an objective viewpoint. Others, however, would argue that teachers should also be involved in role-taking exercises in order to model the activity and to share at least some of the experience of participants.

Dorothy Heathcote's work on improvization is relevant here (Wagner, 1979). In the early stages of a drama, she recommends that the teacher steps forcefully into role in order to harness the energies of the group and get participants involved. Once the students have begun to develop themes in their own way the teacher can play a less prominent part, but at critical stages should step out of role and, as teacher, focus students' attention on issues which have arisen. Thus, teachers both in and out of role can demonstrate sensitivity to the feelings of individual participants (shyness, embarrassment, inhibition), indicate their understanding of aspects of the interaction such as the cohesiveness of the group or conflict among sub-groups, and highlight, when necessary, issues of gender or status.

In order to have the freedom to step in and out of role with ease, Heathcote recommends that the teacher takes a 'middle-rank' position (e.g. first mate, factory supervisor, shop assistant). This not only eases the process of stepping in and out of role, but also keeps open lines of communication with all participants, whether above or below the teacher in the drama; this in turn encourages all students to reflect on their experience and go more deeply into the meaning of what they have done.

[. . .] If group work is about harnessing different perspectives and using the resources of the peer group, then role play and simulation are alternative structures to discussions and problem-solving tasks.

■ Assessing group work

As the new GCSE examinations call for more discussion-based learning and collaborative group work, so one powerful set of constraints on their development will diminish but there will be a need for more definitive guidance on the complex problem of assessing group achievement. [. . .] All we will say here is that there seem to be different criteria for different purposes; some criteria relate to the products of learning, while some criteria relate to the process of learning with teachers increasingly ready to judge the success of group work in terms of such things as engagement, enthusiasm, and participation levels, comparing their pupils' response to group work with their response to more traditional teaching approaches.

Here are some of the dimensions of the problems of assessment.

☐ Short-term and long-term gains

It is sometimes difficult to measure the immediate short-term gains of group work. As we saw in one school, the readiness of pupils in a unit for the 'disruptive' to share their food was regarded as a breakthrough by their tutor, but it did not appear until they were two or three months into the course. Out of context, it might not even be taken into account as a significant, let alone measurable, change in behaviour.

☐ Personal development in the group

Some of the qualities which are facilitated by cooperative work are by their nature hard to quantify. Take another example from the disruptive unit. When pupils rather than the teacher took responsibility for the antisocial behaviour of one individual and showed sensitivity to his needs, they showed the beginnings of an empathy which up to that point had often been lacking in their social relationships. In the teacher's words:

> 'We had a "runner" – somebody who came in one door and went out the other. He was not there one day and they sat down and said that he ought to be kicked out. "Get rid of him. He's no good. He's wasting his place here; he's not doing anything". And we were just about to say, "Is that fair?" when – you get that you sort of hesitate – one of the kids said, "Well I don't know. He's all right when he's doing so and so" and

then the conversation got round to not "we ought to kick him out because he's letting us down" but "what can we do to get him back?" . . . Eventually, they agreed not to nag him when he came back but to say "Oi! You missed something and we've put your name down for this . . ." It's much better to come from them . . . "How can we support somebody because he's one of us?" . . .'

☐ **The assessment of the individual in the group**

We are concerned that although in theory group activity is endorsed in new examination syllabuses, there is often a requirement to identify the contribution of each member of the group and to assess it individually. This in our view contradicts the very essence of group work! However, the issues are real and some teachers would disagree with us because of inequalities, within groups, in contribution. Some pupils opt out by letting the keen members of the group do all the work. Are they to get credit for the end product? A related problem is the tendency of group members to adopt different roles and therefore not to develop skills in areas where they are weak. A science teacher commented:

'It wouldn't worry me if I thought that they took it in turns to play the different roles, but they don't. It always tends to be the same one that will get up and collect the apparatus, do the measuring, do the weighing, whatever it is that they have to do. It will be the same ones that sit and watch, or not even get involved at all.'

How can teachers assess the various levels of participation?

'You can make it difficult by insisting that everyone tries to do an equal share of whatever it is that you are asking them to do. It would be easy to sit back and say that as long as that group came up with some results it doesn't matter provided they can interpret these results and use them. But maybe the watcher gets as much out of it as those using their hands . . .'

Although the contribution of some individuals may be greater than others, it might still be difficult to give due credit to a contribution which represented a considerable achievement for that particular pupil even though in objective terms the contribution to the overall task was not very great. Take the example of Jenny, a withdrawn 15-year-old who opts out of class work and is often tired and listless. During a class debate in which different groups present their own point of view, Jenny twice asks a question. For her this is an achievement although taken in isolation her contribution is not noteworthy.

Some pupils talk a great deal in the group context and seem to be very involved. Yet how do we assess the contribution of Mary who observes silently for a whole lesson until, right at the end, she makes a statement which clarifies the problem and enables the group to make a quick completion of the task?

There are ways in which group productivity and cohesiveness can be monitored, but many teachers see monitoring as likely to undermine the spontaneity of group activity. The awareness of being assessed could be inhibiting. Furthermore, it could act against the very qualities which group work aims to achieve, since individual competitiveness could sabotage group co-operation.

■ Where can group work find a place in the secondary curriculum?

'I can't think of any subject where group work can't be a part of it. All subjects should be doing some group work and it would make it so much easier for those subjects where we are doing it because what you're fighting is these kids wondering what it's all about, thinking, "is this proper work?"'

■ References

Barnes, D. (1976) *From Communication to Curriculum.* Harmondsworth: Penguin.

Bridges, D. (1979) *Education, Democracy and Discussion.* Windsor: NFER.

Galton, M. (1981) Teaching groups in the junior school: a neglected art. *School Organization,* 1 (2), 175–81.

Jaques, D. (1984) *Learning in Groups.* London: Croom Helm.

Stenhouse, L. (1970) The Humanities Curriculum Project: An Introduction. Revised by Rudduck, J. (1983) Norwich: University of East Anglia Press.

Wagner, B.J. (1979) *Dorothy Heathcote: Drama as a Learning Medium.* London: Hutchinson and Co.

5.4

Methods of Teaching STS

J. Solomon

[In this article Joan Solomon takes up the discussion of the teaching methods outlined in the last article in the context of 'science, technology and society' (STS) courses. She examines some of the research evidence on what and how pupils learn through these methods.]

[. . .]

■ Methods of teaching STS

The special behavioural objectives of STS courses have led to some innovations in teaching method. It was clear that the aim to teach about the methods of decision-making within society would be likely to involve classroom strategies which went some way towards mirroring them. Democratic debate in the political arena suggested classroom discussion. This, it was thought, would allow the students to make up their own minds while, at the same time, training them to be attentive to and tolerant of the views of others. So much was agreed by most teachers: the problem was how this should be achieved. Some teachers advocated free discussion right from the start of the course, others that important content should be taught as background to a later but more informed discussion. The stance of the teacher was also problematic. Much was written about the teacher as 'neutral chairman' of class discussion (see, for example, Stenhouse in *The Humanities Curriculum Project*, 1970); later it was suggested that the teacher's job was to present any side of the argument which had been omitted so that a 'balanced view' could be presented. Thirdly the argument that no teacher could, or should, hide their own views began to be heard. If the aim of an STS course was to encourage students to become responsible citizens who would participate in matters of public concern, was it sensible for teachers to pretend that they themselves had no opinion? Traditionalists have often accused STS teachers of indoctrinating the young and, unless teachers respond by representing all education as a kind of indoctrination, it is hard to see how they can defend themselves from the charge.

Gaming, simulations and role play had only just begun to infiltrate the classroom, and were almost unknown in science lessons, before the advent of STS. The earliest appropriate games were often about the location of a power station, for example, Ellingham and Langton (1975). Since that time games and simulations have multiplied and diversified. There is now a great range from simple card games with large elements of fun and luck to more sophisticated and well documented packages where a number of outcomes are possible depending on the judgement of the individual players. With the growth of computer software for schools, it was inevitable that some interactive materials on STS themes should have been produced. The best of these include considerable quantities of accessible data which enable the student to control various factors in a complicated environmental issue. In the larger sense, however, it does seem curious that a solitary activity in front of the VDU should be used in place of interpersonal reactions for teaching the empathic listening to viewpoints of others which is so essential when difficult cultural and social issues are being considered.

☐   Motivation and gender issues

The personal aspects of STS, and also how it is taught, have been claimed by several educationalists (for example, Harding, 1986) to be more comfortable to the way in which girls like to learn about science than is the traditional method. The physical sciences have always had an impersonal image, and psychologists like Head (1980) have suggested that the adolescent boys who choose to study physics often do so because they feel more comfortable following an authoritarian line than having to recognize individual differences and to express personal opinions. Since these are some of the very characteristics which STS courses emphasize, and since factor analyses of girl and boy students often show that person-orientation is the most sharply differentiated characteristic of girls, it does seem likely that STS courses would help to redress the gender balance and attract girl students to the study of science.

But the problem of the motivation to study science goes beyond trying to produce what has been called 'girl-friendly' science. Studies in the United States and elsewhere have shown a steady decline in interest in science and in motivation for studying it during the years of secondary school education, and amongst boys as well as girls. Nor does it grow in popularity with the public at large. Several commentators have attributed this to a presumed uncaring attitude amongst scientists. The research of Fleming which will be described in the next section also supports this view. If this belief about science and scientists is widespread, then more STS education in schools might bring wider benefits than merely adding a liberal seasoning to the school syllabus. It might change the public view of science.

The nature of a subject is not uniquely determined by its subject matter: it is created by the community who practise it and is demonstrated in schools by those who teach it.

'Science is an institution in the world which is progressively presented to the child. The latter creates of science an object to which it then relates. The nature of this object and of the relationship to it will depend on the outcome of nurturing of the child, on the forms in which science is presented and how these interact.' (*Harding, 1986, p. 165*)

The contradictions and dilemmas of this part of the analysis appear in that it is only those courses which aim to develop pupils' civic opinions and action which are attacked for being doctrinaire. And those features which might attract girls to science — openness and capacity for helping people — are at the opposite extreme to those for which adolescent boys so often choose to study it, and how the general public perceive it.

☐ Thinking about social issues in two domains

STS courses have tended to make huge educational claims; with citizenship, decision-making, and personal values at stake they have a great deal to aim for. In her essay on *Teaching about Science, Technology and Society*, McConnell (1982) wrote:

'Public decision making by citizens in a democracy requires an attitude of attentiveness; skills of gaining and using relevant knowledge; values of which one is aware and to which one is committed; and the ability to turn attitudes, skills and values into action. All these steps can be encouraged if a decision making perspective is incorporated into the educational process.' (p. 13)

The time is ripe for some examination of these claims. In particular we shall want to know if STS courses do encourage the kind of citizen attentiveness that McConnell wants. This is the hub of the matter. We would be surprised if knowledge and the skills involved in gaining and using it were not teachable, at least to the more able students. That is the aim of most science courses. Likewise it is not particularly difficult to raise the affective side of problems in school classrooms and observe the values that students express. This is done frequently in religious and social studies. The more penetrating question is what happens when the value and the cognitive systems interact in the context of a science-related issue. Do they produce attentiveness and a better decision-making capability?

Many educationalists, like Mary Donaldson, have argued that children would learn better if the subject matter they were being taught was *embedded* in the thinking of the everyday world rather than being abstracted from it. It is an appealing idea and closely related to the argument about motivation for learning science which was mentioned in the previous section. Unfortunately, research results have not been entirely supportive of it. Henle (1962) showed how difficult it was for graduates to apply logical thinking to problems in which they were closely concerned. The same sort of result was obtained in the science classroom by Dreyfus and Jungwirth (1980) who compared how biology pupils applied logical reasoning in scientific and in everyday situations. It seemed that the affective and value-laden attitudes provoked by the social context made the skills of applying knowledge or logical processes more difficult, especially for the less able pupils.

That evidence is not very surprising but it does highlight the difficulties that education in STS faces. When an issue has already been met and has raised affective judgements, the commonplace or life-world system of thinking, which uses value claims and typifications in place of logical argument and application of knowledge, may become paramount.

Several studies have shown that television may play a large part in informing viewers and raising their value judgements, but the evidence is less straightforward than is often assumed. Whilst it is true that many American studies of high school students (Weisenmayer et al., 1984) have shown that they attribute the greater part of their knowledge of environmental issues to television, the same students are certainly not uniform in their value judgements. Studies on attitudes to television itself (for example, McQuail, 1983) have suggested that viewers interact with the information being presented almost as though it were a social occasion in which they were agreeing or disagreeing with a friend's opinion. Indeed he calls this process 'para-social interaction'. This implies that values will not be taken over wholesale from the media but negotiated through the channels of social or life-world thinking.

Investigations of the different methods of teaching recommended for STS have shown a mixture of results, partly at least because the researchers have looked for increases of enjoyment, of motivation for learning, for value development and for actual learning itself. This is too rich a range of outcomes for easy interpretation. A study of the use of simulation exercises, for example, found them to be not as good as more traditional methods for teaching concepts, but more enjoyable for the students, and more motivating. Teaching for problem-solving skills may increase the students' powers of analysing a situation for either its conceptual content, or its social values, but not necessarily for both. An empirical action research study by Maple (1986), for example, seemed to achieve significant success in teaching the control of experimental variables in school experiments, but could produce amoral travesties of experimental design when students applied their knowledge to societal situations.

These sorts of results, and others like them, have suggested that there are two quite contrasting domains of knowledge. In one the concepts are decontextualized and the mental processes involved are strictly logical. In the other (life-world) domain arguments are conducted about what would happen in a particular context, are expected to be opinionated and evaluative, are socially negotiable, and are not thought to extend to other contexts (Schutz and Luckmann, 1973). Movement from one domain to another is like a cognitive jolt, and is hard to achieve.

There is research on students' views on energy (Solomon, 1985), and also results from practical work conducted by the Assessment of Performance Unit Review Age 15 (DES, 1989) which tends to support this view. In the first study, Grade 8 students who had learnt a course on the physics of energy made richer evaluative judgements on the social uses of energy, when they were making no attempt to use their school learnt knowledge about the energy concept. The abler pupils who did use abstract scientific knowledge about energy seemed to find it harder to bring their evaluative judgements to bear upon the social issues at the same time. It was as though moving from one domain of knowledge to the other was too taxing. Closely related work seemed to show that students even store information acquired from the different domains of knowledge separately from each other (Solomon, 1985). The APU study of experimental work was designed to test for the same practical skills through two different sets of problems. In one the question seemed to be scientific since the context was the reaction times of chemicals. In the other the problem was about the time taken for sweets to dissolve in the mouth. Quite different results were obtained and the students were more, not less, successful when the context was scientific. Perhaps the scientific context contained less of the vague generalizations and affective reactions which might distract the students.

□ **Recommendations for STS education based on research**

Aikenhead (1988) has used a careful multi-method study of Canadian students' views on science, technology and society to point out this divorce between the scientific and the social, and urge a change in science teaching. His Grade 12 students claimed that they had gained most of their knowledge from television and closer questioning showed this to be more from cartoon characters than from more serious programmes. What they did think that they had learnt from school science classes, about the scientific method, was 'almost as inaccurate as the images conveyed by television'. Worst of all,

'. . . the students basically expressed the belief that science and technology have little to do with social problems.'

Aikenhead believes that his study provides backing for the criticism that science instruction is wrong to ignore the social and technological context of authentic science.

Another study of Canadian students, Fleming (1986), used interviews to probe for personal reactions to science-based social issues. He offered the students information booklets to provide essential scientific background knowledge, but these were consistently rejected. Students seemed to have views on what scientists, as people, might think, and this relegated scientific knowledge to the status of personal opinion. We might interpret this as use of the socially acquired life-world attitude, with its empathic understanding, and reliance upon negotiated meanings, for thinking about scientific knowledge. Taking the scientific information in this spirit it is not surprising that the students felt no compulsion to use it in their evaluation of social issues. Like television viewers they treated the information provided as a para-social interaction. At the same time they expressed the view that 'the real facts' would solve all the socioscientific problems. Perhaps we should deduce from this that the students did not recognize any difference between the skills of social evaluation and those of pure cognitive reasoning from unimpeachable premises.

If this were the case, then what pupils learn in conventional science lessons would be received in one of two fundamentally mistaken ways. Either it would be assumed to be no more than a cluster of new and negotiable life-world meanings, or it would be received as 'truth with a capital T' which would, the students claimed, obliterate any personal variations in the evaluation of its social application. In a trenchant criticism of school science, Fleming concludes:

> 'It has often been argued by science educators that the analysis of socioscientific issues requires a background of information. It has, mistakenly, it appears, been assumed that this is scientific data. Instead adolescents require a thorough understanding of the processes of science which generate these data . . .
>
> The perceived obsession of science with the production of facts also allows one to deny any human side to science. Repeatedly, adolescents reported that scientists were interested in progress, and that progress was not concerned with human welfare. . . . Thus science curricula must present science as a product of human endeavour . . . the personal and emotional commitment to the creation of knowledge must be presented.'

In the converse situation, when quality of evaluative, ethical, and moral reasoning is being assessed, researchers often report that the knowledge component is essential to the process. Just as Fleming argued for the human side to the learning of science, Iozzi (1979) has argued for a knowledge foundation to social decision-making. He made a special study of the development of moral reasoning in environmental education and decided that it depended on background knowledge, as well as on interest and concern. He argued that both of these kinds of factors must be present in the learning process if decision-making is to be achieved.

There is precious little research on the results of STS education within school. In Holland the PLON project for teaching physics with special emphasis on its social aspects has been in operation since 1980. Eijkelhof (1985) reported the results of a small pre- and post-test study of some Dutch students studying a PLON unit on ionizing radiation which emphasized the risk to health from nuclear and X-rays, and encouraged discussion of its acceptability.

The unit engaged the students' interests well, but changed their attitudes to radiation very little. In particular the issue of radiation from nuclear waste was assessed almost exactly the same as it was before the course, and with the use of the same kinds of common-sense arguments. On the other hand a question about using radiation for the preservation of food elicited more favourable responses after the course, and many of the students showed some valuable use of knowledge derived from the course.

Eijkelhof speculates that the students had already made up their minds about nuclear radiation from power stations before the course began since the subject has had high visibility in Holland for many years. How early adolescents begin to consider topical concerns in a personally committed sense is very hard to know. The review article of Weisenmayer et al. (1984) suggests that environmental attitudes are formed very early, often before the students reach Grade 8 (12-years-old), and are then very resistant to change. Further inputs of information are welcomed, but tend to do no more than polarize existing views. The students seize upon facets of the ideas presented that support their own views and then tend to ignore the rest as irrelevant or biased. This is clearly not the outcome of STS education expected by McConnell.

It would be pleasant and satisfying if this [article] could end with recommendations for STS within the curriculum which were supported both by empirical research and by educational polemic. Unfortunately, the dilemmas which have dogged the implementation of STS also plague research into its school operation. It appears, as far as we now know, that students' attitudes are strongly influenced by out-of-school factors, and that they do not easily use the scientific knowledge which we teach them in conjunction with personal evaluation for social decision-making. Holistic educational theory would insist that knowledge and evaluation are complementary and essential characteristics of human development, but offers no advice on how they should be taught.

But some of the research data can be used to bring the argument full circle. If, as Aikenhead and Fleming insist, the students do not perceive the difference between socially acquired negotiable knowledge, and abstract overarching scientific knowledge, then it is the first approach to STS (through an understanding of science as a way of knowing) which offers the most promise. Perhaps a philosophical introduction — appropriate to the level of the students — is a prerequisite for freeing the reasoning and valuing faculties to work together in social decision-making. Only more research can tell.

■ References

Aikenhead, G. (1988) Student belief about science–technology–society: four different modes of assessment, and sources of students' viewpoints. *Journal of Research Science Teaching*, in press.

Department of Education and Science (1989) *Science in Schools Age 13. Review Report.* London: HMSO.

Dreyfuss, A. and Jungwirth, E. (1980) A comparison of the 'prompting effect' of out-of-school with that of in-school contexts on certain aspects of critical thinking. *European Journal of Science Education*, 2 (3), 301–10.

Ellingham, H. and Langton, N. (1975) The power station game. *Physics Education*, 10, 445–7.

Eijkelhof, H. (1985) Dealing with acceptable risk in science education. Paper presented to the conference 'Science Education and Future Human Needs', Bangalore.

Fleming, R. (1986) Adolescent reasoning in socio-scientific issues, Part I. *Journal of Research Science Teaching*, 23 (8), 677–88.

Harding, J. (1986) The making of a scientist? In J. Harding (ed.) *Perspectives of Gender and Science*. Lewes: Falmer Press.

Head, J. (1980) A model to link personality to a preference for science. *European Journal of Science Education*, 2, 295–300.

Henle, M. (1962) The relationship between logic and thinking. *Psychological Review*, 69, 366–78.

Iozzi, L. (1979) Moral judgment, verbal ability, logical reasoning ability, and environmental issues. Unpublished PhD thesis, Rutgers University, NJ.

McConnell, M.C. (1982) Teaching about science, technology and society at the secondary school level in the United States. *Studies in Science Education*, 9, 1–32.

McQuail, D. (1983) *Mass Communication Theory: An Introduction*. London: Sage Publications.

Maple, J. (1986) An investigation of the transferability of practical skills. Unpublished MSc thesis, University of Oxford, Oxford.

Schutz, A. and Luckman, T. (1973) *The Structure of Life World*. Evanston, IL: Northwestern University Press.

Solomon, J. (1983) Eight titles in the *SISCON in Schools* series and a Teachers' Guide. Hatfield: Association for Science Education and Oxford: Basil Blackwell.

Solomon, J. (1985) Learning and evaluation: A study of school children's views on the social uses of energy. *Social Studies of Science*, 15 (2), 343–71.

Stenhouse, L. (1970) The Humanities Curriculum Project. London: Heinemann.

Weisenmayer, R.L., Murrin, M.A. and Tomera, A.N. (1984) Environmental educational research related to issue awareness. In *Monographs in Environmental Education and Environmental Studies (1971–1982)*, pp. 61–94. ERIC, Ohio State University.

5.5

Production Simulations

I. Jamieson, A. Miller and A.G. Watts

■ **Introduction**

Production simulations are attempts to mirror the management and organization of production in manufacturing industry. They illustrate a common work process and show the principle of specialization in action. The fact that the concept of the division of labour is a common theme in many economics and other courses helps to explain the popularity of production simulations in schools.

As work simulations, production simulations are flexible in that they can be used to give experience of various work roles, including those of supervisor, quality controller, shop steward and assembly-line worker. In addition, they provide opportunities for students to practise management skills, particularly in motivating the workforce and in organizing the smooth conversion of inputs of resources into outputs of finished goods. The presence of management and organized labour means that industrial relations, including negotiations on pay, conditions of work, and health and safety, are a frequent feature. More sophisticated versions also reproduce key features of the work environment, such as clocking in, wearing industrial clothing, and taking breaks in a works canteen: these can be termed 'factory simulations'.

Production simulations, therefore, vary from relatively simple activities with no pre-specified work roles, to complex factory simulations which attempt to mirror the key features of factory life. An illustrative example appears in Box 1 (for a case study of a similar Lego simulation, see Quantick, 1985).

■ **Key variables**

There is a range of published production simulations which are suitable for use in schools. Compared with business games they are also relatively easy for teachers to construct, and there are several examples of locally developed

Box 1   An illustrative example

A group of 60 lower-sixth-formers are involved in an Industry course. One morning session is devoted to a production simulation involving the manufacture of toy cars out of Lego bricks. To familiarize the students with the process of assembling the cars, each student is given a box containing parts and pictorial instructions. They are asked to time themselves to see how long it takes to assemble the complete car. Once these times are recorded on a chart, the students form themselves into groups of four. Each group is asked to assemble five cars and again to time how long this takes. Groups consider how this task can best be achieved, some groups opting for a production line with each student assembling particular parts only. The average time taken to build each car is recorded and the teacher makes the point that the division of labour should lead to increased efficiency.

At this stage, a manager and trade unionist from a local firm are introduced to the students. The students are divided into four groups of 15, and each group or 'factory' is asked to choose a general manager to be in overall charge. The management team is completed when the four general managers select their works managers and supervisors. The management teams retire to another room with the local manager to be briefed upon their role. They discuss the organization of production, training the workforce, and company accounts. Meanwhile, the remainder of the workforces are electing shop stewards who are briefed on their role by the trade unionist.

Once the management team re-enters the room, they start to organize their production line, shifting both people and furniture until they are satisfied that the process is smooth. Individual workers are briefed on their jobs, including a quality controller who must check the accuracy of the finished product and the stores from where new parts are obtained (the stores are also used to break down the cars in order to ensure a continuous flow of parts). The local industrialists are on hand to act as advisers if their help is requested.

The teacher initiates the first round of production by giving each factory an order for 15 cars in 10 minutes' time. Each line bursts into action, but problems soon surface. Bottle-necks appear as some workers cannot keep up with the pace of the line. At the end of the round, it is clear that two factories have not met their orders. Managers do their accounts, adding up their costs and deducting them from the sales revenue derived from the cars they have produced. At the same time, the shop stewards discuss pay and conditions with the assembly-line workers. Many are dissatisfied with their pay, given the hectic pace at which they are being asked to work. In the ensuing bargaining session, the shop stewards attempt to wring concessions from the management.

The local industrialists observe these interactions. New orders are placed and the next round of production begins. The output and profit positions of the factories are posted for all to see. The simulation continues for four rounds until serious industrial relations problems occur with stoppages of work on two of the lines.

Debriefing begins in the factory groups, students engaging in an exchange of views and feelings about each others' roles and actions. The whole group is brought together, and the manager and the shop steward are asked to comment upon the performance of their student counterparts in the simulation. They discuss the parallels and contrasts between the Lego factories and their own workplace, in which production seems more continuous and less strife-ridden. Various points are made about the organization of production and the conduct of industrial relations.

materials, some of which are described later in this [article]. The key variables on which production simulations can be distinguished from one another are:

- *Origin*   Some are based directly on published materials, some are adapted from published simulations, and some are teacher-devised.

- *Nature of simulation*   The most widely used production simulations are manually operated. Some, however, are either computer-assisted (i.e. using the computer as an aid) or computer-based (i.e. played on a computer).

- *Key objectives*   The designers of published simulations usually specify the educational objectives of the activities. These may include: understanding the world of work, particularly the production process and work roles in production; industrial relations; human relations at work; and economic and industrial concepts such as fixed costs, the division of labour and collective bargaining.

- *Functional areas covered*   Although the prime focus is on the production function, other areas of business can be built in − for example, finance, personnel, administration, and marketing.

- *Production mode*   Some simulate small-batch production, whereas others mirror mass-production methods.

- *Nature of the product*   The product is frequently a simulated product made from paper − for example, paper chains or bridges − or Lego bricks. In locally developed simulations, actual products might be assembled or manufactured − for example, Christmas crackers or toy nurses' uniforms (see pp. 251–9).

- *Degree of student autonomy*   Some simulations specify the details of the production process, while others allow student managers to design the most efficient method of production.

- *Worker roles*   Most production simulations have a minimum of two work roles − workers and managers − although many include other roles too (see Table 2, p. 263).

- *Organizational features*   Production simulations are used in various curriculum contexts, ranging from CDT to social studies. Time is a key variable. Complex factory simulations making real products can take days, whereas simple assembly-line activities and some computer-based production simulations can be fitted into a double period. Most production simulations have been designed with the 14–19 age-group in mind, although many can be adapted for use with young pupils.

■ Four published production simulations

The differences between production simulations can be illustrated in more detail by describing four published activities.

(1)   ASSEMBLY LINE. This is a production simulation which seeks to provide 'a dramatic and dynamic model of a mass production system'. It aims to recreate the conditions in which students can experience the human problems associated with mass production: alienation from the final product; the monotony of repetitive work; the pressures to speed up or slow down production; and the stresses of factory life. Although the simulation has been most widely used in the economics curriculum in the upper school, it can also be used with younger pupils, including first-years (Nuttall, 1986).

   The production is a paper vintage car which has been printed on a piece of A4 paper. Students have to cut out, colour and stick together the different parts to make the finished vehicle. A conveyor belt is simulated by placing a continuous piece of lining paper along a row of desks. The teacher begins the simulation by placing orders for a number of cars to be produced within a given time-period. The production engineers organize the line and train the workers while they are 'on-the-job'. Supervisors iron out production problems by checking faulty parts, supplying materials where required, and reassigning workers to help the flow of cars along the line. Inspection workers provide quality control, while the maintenance crew check the 'belt' and clear away rubbish from the assembly line.

(2)   THE BRADFORD GAME. This is one of the most widely used production simulations. Students work in teams which represent companies in the same market for paper notebooks. Each group is given the same basic raw materials (paper) and equipment (scissors, rulers, staplers, punch, etc.) with which to manufacture the notebooks. Companies organize the production process and aim to make a profit by completing customers' orders. The controller acts as a wholesaler from whom additional resources can be bought, and as a customer who places orders for notebooks with varying specifications (size, colour of cover, number of holes). Quality control can be carried out by the companies' customer(s), who can reject sub-standard notebooks. The 'game' element is provided by the competition between companies to make the most profit. The flexibility of THE BRADFORD GAME is illustrated in Box 2.

(3)   WORKERS AND MACHINES. This is a computer-based production simulation which provides a 'dynamic visual representation of a production process'. Students take on the role of production supervisors who must try to maximize output and minimize unit costs on an assembly line producing supermarket trolleys. The simulation is very flexible in that it can be played individually, or in groups working on the computer, or as an inter-team competition. WORKERS AND MACHINES resembles a business game in that it is played over ten rounds or 'shifts': each round involves the making of decisions and receiving feedback on results.

Box 2   Some alternative uses of THE BRADFORD GAME

A modified version of THE BRADFORD GAME is extensively used in the London Borough of Haringey in the context of schools-industry work. Its popularity is attributed to several factors: the paper notebooks have a subsequent use which provides a *real-life referent*; *specifications* can be varied to emphasize different teaching points and to maintain team alertness; *quality control* is an important concept in the game; *adults other than teachers* can participate in the roles of wholesalers, bankers or customers; and the game can be played *cooperatively*. THE BRADFORD GAME is used in Haringey in three main ways:

- *As preparation for a mini-enterprise*   The simulation introduces several concepts and techniques which are useful for students to encounter prior to taking part in mini-enterprises, including working in a team, the division of labour, assembly lines, the purchase and use of raw materials, the importance of product quality, and simple accounts. A cooperative version of a mini-enterprise can be preceded by a co-operative approach in THE BRADFORD GAME. Thus students are permitted to practise production prior to the start of the game, and no pressure is placed on the cooperatives until they can make a good product to concrete specifications. The debriefing focuses on ideas that will help students in their mini-enterprises.

- *As part of fifth- and sixth-year core courses*   The main objectives of using the game with these students is to provide simulated work experience, to illustrate the organization of production lines, and to demonstrate management in action. Managers are briefed separately, and must allocate roles and brief their workers. This model leads to industrial relations problems, and conflict between management and workers generally predominates over production.

- *In an Industry module*   THE BRADFORD GAME has been played over a period of a half-term during an Industry module. Initially the focus is upon production problems, but this gives way to the con-sideration of wider questions related to local employment issues. For example, the impact of cheap imports is simulated in the game, and parallels are drawn with the history of the furniture industry which suffered as a result of similar competition.

The decisions which the production supervisors have to make include the allocation of a force of 30 workers to the three operations of cutting, shaping and welding. The scheduling of the inputs of raw materials takes place between rounds, and a visual display indicates the supply of unused raw material and the number of trolleys at each stage of production. A 'workers idle' chart shows the proportion of labour time that has been wasted; in addition, feedback is given on labour productivity and unit costs. The computer program allows the teacher to introduce random factors which affect at least one of the three processes: for example, absenteeism, and new technology (the program enables the supervisors to assess its impact upon the production process). The simulation is designed to teach the efficient use of resources and the law of diminishing returns. It illustrates that new technology can raise output, but also that it can alter the pattern of demand for labour and can require greater labour mobility between jobs.

(4) CHAIN GANG. This is a computer-assisted production simulation. It also contains features of business games (inter-team competition and computer print-outs of results) and of negotiation simulations (separate wage-bargaining sessions beween rounds). It is designed for use with a wide ability range of students aged from 14 to 18. The tutor's manual contains substantial supporting material on industrial relations for use in preparatory work.

In the simulation, students take on the roles of managers and workers in competing companies making paper chains to precise specifications. The principal aim of the managers is to maximize the company's cash-in-hand at the bank. The workers' main objective, operating through their elected shop steward, is to achieve a level of earning comparable with workers in other companies. Each chain factory has a manager who is provided with the requisite materials: paper, a ruler, a pencil, scissors, a stapler, and a sample paper chain. The manager begins by organizing the production process and allocating functions among the workforce. Production occurs during three fifteen-minute rounds, during which companies try to complete the orders they have been given by the controller. At the end of each round, a negotiation phase takes place where wage claims and offers are discussed. If the manager and shop steward cannot agree, the controller acts as an arbitrator, but he or she must decide in favour of the side with the best arguments. Between rounds, companies receive print-outs which show their financial position, particularly the cash flow and the production record.

■ Case study: CHAIN GANG in action

CHAIN GANG was used with a group of twelve students approaching the end of a BTEC General course at Willesden High School in Brent. The mixed group of nine girls and three boys had experienced various work simulations during the course, and were keen to experience more activities of this kind. The teacher felt that CHAIN GANG would provide a way of integrating and applying learning from the different strands of the course. It was also hoped that two girls who had just returned from work experience would be able to apply some of their learning in the simulation. They were accordingly allocated management roles.

The simulation was held in adjoining rooms at the local teachers' centre. One room was laid out with chairs in a circle for briefing and debriefing; the other room had chairs grouped around three tables, one for each team. There was a number of advisers on hand throughout the day, including a production director and a union official from the same manufacturing company, a representative from the local trades council, and an industrial training officer.

Prior to the arrival of the students, the advisers were briefed as to their role during each stage of the simulation. The student teams were pre-selected by the teacher: the criteria for selection included ensuring a 'natural leader' in each team, distributing the three boys across the teams, and having a mix of ability. Although the students were told that managers had been selected at random, this was engineered so that what the teachers described as 'forceful' individuals would become the managers.

The students were told that the simulation would raise issues about relationships between management and workers. The managers then went into the 'production room' for a separate briefing with the management advisers. They were talked through their brief and were given several minutes to examine the contents of their company's resources box. During this time, they had to plan the production process and to consult with the advisers. Meanwhile, the workforce was being briefed by the union representatives. Overall, little attention was given to competition between the chain factories: the criterion for success was maximizing cash-in-hand.

When the activity started, workers applied themselves to the task of making their chains. One adviser observed that the management style was noticeably 'teacher-like', adding that this was unremarkable given that teachers were a prominent role-model for students. The managers helped the workers to speed up production by reducing bottle-necks on the production lines. They began to show signs of exasperation at the failure to achieve production targets. Nevertheless, little industrial conflict was evident in the companies, each of which functioned well as a team. One manager commented that there was a 'lot of support from everyone', while a worker said of her manager that 'she pushed you a lot and was a very good manager: we pulled together to make up time lost'.

After three rounds of the simulation, the teams were asked to report back on the following questions:

- How was production organized?
- How well do you think you functioned as a group?
- Were production targets/orders met? If not, why not?

In all cases, it was the managers who reported back to the whole group in the debriefing room. The advisers then commented upon what the managers had said. The trade unionist pointed out that the company which had paid the highest wages had been the most profitable, in contrast to much popular belief. In a final round of evaluation, all the students and advisers were asked to make statements about the value of the exercise. Many students felt that the simulation had been good for encouraging cooperative teamwork. Some of the advisers felt that the industrial-relations side of the simulation had not been developed because of this emphasis on teamwork. The trade unionist felt that management had dictated what should happen, including determining levels of output without adequate negotiation.

■ Some factory simulations

'Factory simulations' attempt to simulate the work environment in addition to the work roles and work processes commonly found in other production simulations (such as THE BRADFORD GAME). This was done in a factory simulation in Coventry (Shapcott and Wright, 1976), which formed an integral and assessed part of a fifth-year leavers' course. Students were prepared for the simulation in their normal curriculum subjects: for example, English included work on job applications, mathematics lessons on wage rates and production graphs, and social studies on work concepts and interviews. But the most interesting feature of the simulation, which involved the assembly and packaging of Christmas crackers on a production line, was the variety of respects in which attempts were made to replicate a work environment:

> 'The factory was sited in a detached temporary classroom which was ideal because the noise from the factory disturbed no-one and it was close to the domestic science rooms being used as the works canteen. To help the factory atmosphere the pupils involved wore 'civilian clothes', used a separate school entrance, worked different hours to the school, clocked in and out, were paid an hourly wage, were stopped money for late arrival or poor work, ate in their own canteen.'

Unlike many shorter production simulations, work roles were rotated so that all the students could experience different production tasks, as well as supervisory, managerial and trade-union roles. This failed to prevent boredom setting in over the two weeks of the simulation. Nonetheless, the rotation of roles stimulated discussion during debriefing on, for example, the impact of different management styles upon output and quality. Assessment took the form of an 'in-depth' profile of each student compiled by the teachers for subsequent use in references.

Another production simulation which devoted a lot of attention to simulating a work environment was a factory simulation at Hetton School in Sunderland (Smith, 1985). The main purpose of the simulation, indeed, was to create a realistic factory environment in which relationships, communications and industrial relations issues could be examined. Two factories making 'real' products of saleable quality were established, with a wide range of work roles available to the participants. Such elaborate production simulations come close to mini-enterprises, except that the emphasis is on manufacture rather than on marketing and selling, and the organization tends to be highly teacher-centred.

The most interesting feature of Hetton Pine Products, however, was the ways in which linkages with reality were created. Industrial advisers were involved in planning and implementing the simulation. Industrial managers interviewed applicants for the post of student manager and subsequently briefed the successful applicant. Visits prior to the simulation gave students a feel of the real atmosphere of a factory. After the simulation, students visited their industrial counterparts at their workplaces, and shadowed them in some cases. These activities highlighted the desirability in a factory simulation of building in reality-linked pressures through the use of costing procedures, of production targets and of a requirement to 'sell' the finished product.

■ Some production simulations in special education

Production simulations have found particular favour in special education, partly because they can offer simulated work experience and at the same time help with basic skills. A group of school-leavers from a special school in North London took part in a production simulation as part of a course including industrial visits, classwork and work experience (Paine, 1983). The simulation arose from a link with a local toy manufacturer, who provided a supervisor to brief the students on the work of assembling and boxing 1000 toy nurses' uniforms for children. It was argued that the students benefited from several facets of the experience, including handling real products, learning to cope with repetitive tasks, needing to keep accurate time, and working co-operatively. In addition, the simulation allowed basic skills to be used in practical situations; for example, numeracy (in stock control and costings), and communication and social skills.

The use of real products in work simulations was illustrated in a South London special school in which a bed was assembled by a group of students. This assembly-line simulation, carried out on an employer's premises, was the culmination of a factory visit to observe the processes involved.

Another case study of a special-school industrial simulation involving production lines lasted for a week at a school in Brent (see Peffers, 1985). This example illustrated how elaborate tailor-made production simulations can be constructed to meet the needs of individual schools. The preparation involved training sessions for students in their work tasks, and playing the LEGO CARS simulation to give them experience of a production line with a simulated product. An interesting element of context fidelity was provided by the circulation of job descriptions, asking students to write letters of application and to attend interviews for jobs in what was termed 'Glenwood Enterprises plc'. Students were then given an employment contract specifying the (small) amount of money they would earn in wages as well as the penalties for late arrival at work. The use of wage payments was designed to add realism and motivation, although all the workers were paid the same amount. Another interesting feature of this simulation was the 'holding company' organization, with an administrative centre using real business documents provided by a local company, and a number of subsidiary production companies making paper booklets, decorative paper flowers and 'social sight' signs ('exit', 'danger', etc.), and a canteen providing refreshments.

■ Issues

Among the issues that arise from these and other case studies are those related to the *role of the teacher*, *progression*, *work roles*, *work processes* and *industrial relations*. We will examine each of these in turn.

□ Role of the teacher

It seems clear that production simulations pose particular organizational problems for teachers. Materials and equipment have to be provided, and the classroom geography has to be reorganized for the production lines. In the CHAIN GANG case study, additional attention was needed to setting up the computer and printer. The degree of planning required is even greater in the case of factory simulations, where large quantities of materials and equipment may be required.

During the activity, there are a range of functions for the teacher to fulfil. As with business games, many production simulations have to be managed because play occurs in discrete rounds. The teacher may also be required to act as a customer buying the finished article, as a quality controller checking the product, as a supplier of raw material (as in THE BRADFORD GAME), or as an arbitrator (as in CHAIN GANG). Often the teacher will intervene in the action to feed in problems which stimulate industrial relations problems. Even where this is not the case, the generation of strong feelings on the line stemming from worker boredom and management high-handedness can lead to industrial action. This adds a risky dimension to the activity, particularly as such breakdowns are so unpredictable. When they do occur, judgements must be made about when to intervene to break a deadlock or to end the simulation.

The teacher also needs to ensure that the simulation is thoroughly debriefed. Production simulations provide an experience of only one industrial workplace, and there is a danger that students will use this experience as a basis from which to draw misleading generalizations. A sample of comments from fourth-year students who had participated in the LEGO CARS simulation illustrates this problem:

'I learnt that [in] production which ever part you work on is boring and tiring.'

'The shop steward hardly does any work he's always at meetings.'

'What surprised me was the power the union had over the production.'

The involvement of trade unionists and industrialists in debriefing is essential if such opinions are to be questioned and explored. The danger remains, however, that a powerful simulation experience is likely to create a stronger and longer-lasting impression than a debriefing session in large groups at the end of a long and tiring day. This reinforces the point that production simulations, like all work simulations, are best employed in the context of a course which provides other forms of evidence about industry and the world of work. [. . .]

☐ **Progression**

Production simulations have been developed for use in very varied settings, ranging from infant schools (Coppack and Shaw, 1986) to providing simulated work experience in school-based factories for fifth-year students in secondary schools. The practical nature of production simulations has made them popular with primary teachers. Often an activity such as LEGO CARS is adapted for use with younger pupils. The fact that the same simulation can be used with a broad age spectrum raises the issue of progression. Table 1 illustrates how progression could be built into the curriculum through a key focus on a particular aspect of the world of work. Thus with infants, the focus might be upon the *task*, whereas with juniors there might in addition be a focus on work *roles*. As conceptual understanding develops during secondary schooling, the focus could shift to work *processes* and eventually, by the age of 16, to work *environments*. In the sixth form, the focus might return to work *processes*, but exploring them at a more complex level than in the lower secondary school. Table 1 elaborates this structure by offering examples of key concepts that might be developed with each age-group through production simulations. [. . .]

Such models might also incorporate other forms of work simulation with which production simulations overlap. 'Design and make' activities become production simulations when work roles are established and the production of a prototype is expanded to include the manufacture of a quantity of products on an assembly line. At the other end of the spectrum, production simulations can incorporate elements of marketing and, therefore, border on being mini-enterprises. Usually, however, the products which are produced are only hypothetically 'sold' before being given to students or deployed in other ways. Production simulations also share many of the features of work practice units, except that they are generally based in ordinary classrooms. [. . .]

☐ **Work roles**

Production simulations offer students the experience of a variety of roles found in manufacturing industry. In addition to shopfloor roles such as line worker, supervisor, and stock controller, staff roles and ancillary roles are frequently

Table 1   A possible model for progression in production simulations.

Age	Key focus	Use and key concepts
5–8	Work tasks	Production simulation used as vehicle for practical work, e.g. cutting, drawing, measuring, and language and number work. *Key concepts*: factory, industry.
8–11	Work roles	Production simulation illustrates different jobs at work. *Key concepts*: manager, supervisor, shop steward, worker.
11–14	Work processes	Production simulation demonstrates basic economic processes. *Key concepts*: division of labour, specialization, factors of production.
14–16	Work environments	Factory simulations provide a simulated work experience using props, 'real' time, actual products, complex role relationships. *Key concepts*: output, efficiency, wages, deductions, fixed and variable costs, management.
16+	Work processes	Production simulation used to explore industrial relations processes. *Key concepts*: collective and plant bargaining, negotiation, procedure, management. *or* Student-centred approach where groups decide on product and process and motivate other groups of 'workers'. *Key concepts*: management style, job design, motivation, job satisfaction.

found in factory simulations. The work roles in the simulations referred to in the earlier part of the [article] are shown in Table 2.

Management has a key role to play in the successful establishment of the production line. Similarly, shop stewards are important when industrial relations issues are a major objective. This raises a more general issue. The temptation to appoint to certain key roles students who the teacher feels are likely to be more effective in these roles may be irresistible. But it is important to question the criteria of selection and to attempt, during the course of a programme of work simulations, to ensure that all students have the opportunity to experience management roles in particular. Although management roles can be stressful, they can also be challenging, can call upon and develop a variety of skills, and can often involve greater contact with visiting advisers.

Students occupying different roles are likely to have very different

Table 2 Work roles in some production simulations.

	ASSEMBLY LINE	THE BRADFORD GAME	CHAIN GANG	WORKERS AND MACHINES	LEGO CARS	Glenwood Enterprises plc	Hetton Pine Products
Main roles							
Factory/production manager	•		•	•	•	•	•
Supervisor	•				•	•	•
Line worker	•	•	•		•	•	
Quality controller	•	•			•	•	•
Stores/stock controller					•	•	•
Despatch						•	
Shop steward			•		•	•	•
Maintenance	•						
Staff roles							
Managing director/ general manager		•	•		•	•	
Secretary/typist						•	
Clerk						•	
Accountant						•	
Ancillary roles							
Designer/graphics							•
Computer operator							•
Journalist						•	•
Canteen worker						•	•
Bank manager						•	
Customer/client	•						

experiences during the simulation. Thus assembly-line workers, whose role within production simulations frequently involves the mechanical repetition of manual tasks, can easily become bored and alienated. This may well mirror reality quite closely, and ought to lead to a critical discussion of assembly-line methods of production. In the 'real world', however, production-line workers tend to be motivated by pay and other benefits which can to some degree compensate for the repetitive nature of the work. Such extrinsic rewards are usually much weaker in production simulations, and this can be a problem, especially where the simulations are of long duration. In contrast, the management and union roles are at the centre of the action, and students occupying them commonly report greater interest and learning than do those acting as line workers. This suggests that there may be a case for rotating roles where possible so that all students can experience worker, union and managerial roles during the course of the simulation. [. . .]

☐ Work processes

The process of production can be prescribed by the teacher, including the layout of the line and the designation of work roles. On the other hand, a more student-centred approach permits the management with or without consultation to design the organization of production: in this way, alternative methods of production may emerge, such as work teams operating on a 'Volvo' model (where semi-autonomous teams of workers build almost the entire motor car). These alternatives can be highlighted and discussed at the debriefing stage.

Elgood (1981) describes a student-centred way of organizing production or 'practical' simulations (as he calls them). Materials are supplied from which a range of simple products can be produced – for example, Christmas decorations. In teams, a product and a production process are designed, and each team then acts in turn as a management team, using the other teams as workers. In this model there is a strong emphasis on negotiation: of materials prices and retail prices with suppliers and customers respectively (the teacher in role), of loans with the banker (again, the teacher in role), and of wages and conditions with the workforce. Through such means, production simulations can provide considerable scope for student autonomy in decision-making.

☐ Industrial relations

The presence of managers and workers adds an industrial relations dimension to production simulations, and developing this dimension is important if stereotyped images of industry are to be challenged. Often negotiations over substantive issues such as pay and bonus payments are an integral feature of each round, as is the case in CHAIN GANG. Such negotiations frequently involve interpersonal bargaining between a manager and a shop steward rather than collective or plant bargaining involving groups of representatives on each side. Usually only one union is involved and all workers are members of it.

Industrial conflicts resulting in stoppages of work are indeed a common feature of many production simulations. This results, in part, from the absence of procedures for settling disputes. The costs of work stoppages for student workers, such as loss of earnings, are often outweighed by the benefits, such as the break it provides from the tedium of working on the line and the excitement generated. The causes compare fairly closely with the major causes of stoppages of work in the UK as set out in Table 3. As is the case in real life, disputes over pay are the most common cause of stoppages in production simulations, as pay negotiations are often a built-in feature. Work allocation can also lead to problems for student production managers who demand flexible working practices from their workforce. Management high-handedness and insensitive supervision can lead to industrial action if it provokes worker indiscipline. In contrast, issues of redundancy, duration and hours worked, and trade-union matters, all figure less frequently.

Industrial relations issues can be a feature of each production period round. Alternatively, a problem can be fed into the simulation at an appropriate moment by the controller. For example, management might be instructed to cut costs, or shop stewards may be informed of some health and safety problem on the line. As we have pointed out, though, the generation of strong feelings in the simulation can sometimes lead to more unpredictable forms of industrial action. To cope with these problems, students can be introduced to disputes procedures during the preparation and briefing. This means that when difficulties occur in the simulation, the 'workers' are clear about what courses of action are open to them. Without such procedures the simulation can end in deadlock much earlier than anticipated by the teacher.

The extent of industrial conflict in production simulations will be influenced by several factors. In the CHAIN GANG case study, the students interpreted the simulation, in the light of their previous simulation and course experience, as a team-building activity. The teacher and industrial advisers, on the other hand, regarded it as a simulation designed to raise issues about industrial relations. But although this aspect was emphasized during the briefing, conflict failed to materialize. Other factors which contributed to this lack of conflict were the small size of the companies (which made the management task easier), the lack of stress on inter-company competition, the

Table 3 Causes of stoppages of work during first six months of 1986.

Cause	Number of stoppages
Pay	171
Manning [sic] and work allocation	89
Working conditions and supervision	57
Redundancy questions	44
Dismissal and discipline	41
Trade-union matters	26
Duration and hours worked	23
	451

Source: *Employment Gazette*, September 1986.

absence of interference by the teachers to stimulate industrial conflict, and the choice of strong 'natural leaders' as managers. The latter may have ensured a smooth production process, but appointing such students to serve as shop stewards might have led to more bargaining.

■ Conclusion

The production line symbolizes what many people think of as British industry. The factory was at the heart of the Industrial Revolution, and its processes have been the major definer of work for two centuries. It is ironic that schools are trying to pay attention to these processes at a time when the organization of work is undergoing major transformations: the result is to leave schools modelling processes that are becoming industrially outdated. Manufacturing has been in decline in Britain for some time, and now employs less than a quarter of the working population (MSC, 1986). Moreover, manufacturing has always exhibited more variety than schools seem to realize. This is true even of the small-batch manufacturing that is normally mirrored by production simulations: thus a production line organized to produce a standard product at high volume for sale ex-stock at a low price will look quite different from one organized to produce high quality and high reliability, perhaps incorporating customer-specified modifications, and selling at a premium price. The advent of new technology is now having a major impact on these manufacturing processes, with the introduction of computer-aided manufacturing and robotics leading, step by step, towards the introduction of fully automated flexible manufacturing systems (FMS). In this situation, what is the rationale for simulations which, in many cases, mirror Henry Ford's first production line for the Model T Ford?

In our view, the key to answering this question lies in the way in which the simulations are used. If production simulations are seen essentially as vehicles for teaching certain enduring elements of the manufacturing process, they may continue to have a useful and defensible role to play. Production simulations are potentially powerful tools for teaching about industrial relations, about interpersonal relations at work, about the management of people, and about the economics of cost control. Such factors are arguably essential features of any production process, whatever the technology. Schools will then have to find other ways of making students aware of the growing technological gap between the production processes they are experiencing and the emerging production processes in the modern industrial world.

■ References

Coppack, K. and Shaw, E. (1986) Infant industry: a production simulation. *SCIP News*, No. 14.

Elgood, C. (1981) *Handbook of Management Games*. Aldershot: Gower.
Manpower Services Commission (MSC) (1986) *Labour Market Quarterly Report*, June.
Nuttall, P. (1986) Division of labour: parents in the classroom. *SCIP News*, No. 14.
Paine, E. (1983) 1,000 Nurses' Uniforms. *SCIP News*, No. 7.
Peffers, J. (1985) An industrial simulation at Woodfield Special School, Brent. In I.M. Jamieson. (ed.) *Industry in Education: Developments and Case-Studies*. Harlow: Longman.
Quantick, P. (1985) Kingsdown Cars: an industrial simulation at Kingsdown School, Wiltshire. In I.M. Jamieson (ed.) *Industry in Education: Developments and Case-Studies*. Harlow: Longman.
Shapcott, D. and Wright, V. (1976) The Christmas Cracker Factory. *Careers Adviser*, 5 (4).
Smith, K. (1985) Alternatives to work experience at Hetton School, Sunderland. In I.M. Jamieson. (ed.) *Industry in Education: Developments and Case-Studies*. Harlow: Longman.

■ Published simulations referred to in the text

ASSEMBLY LINE (production simulation). Developed by Dobbs, D., Goodell, C.G. and Hill, R.F. In D.J. Whitehead (ed.) (1980) *Handbook for Economics Teachers*. London: Heinemann.
THE BRADFORD GAME (production simulation). Part of *The Production Programme*. (1978) Cambridge: CRAC/Hobsons.
CHAIN GANG (production simulation). Developed by the School Curriculum Industry Project BBC(B) disc. (1985) Cambridge: Cambridge University Press.
WORKERS AND MACHINES (production simulation). Developed by the Computers in the Curriculum project. Understanding Economic Software series. BBC and RML Network disc. (1985) Harlow: Longman Micro-Software.

5.6

Halfway There: Reflections on Introducing Design and Technology into the Secondary Phase

M.E. Harrison

■ Introduction

The phased introduction of design and technology (D&T) into the England and Wales National Curriculum means that this article (written in April 1992) is an interim account of a change process that will continue well into the 1990s. In a typical secondary comprehensive school whose intake is at Year 7 (11+), the current (1991/92) Year 7s are the second such year to have been taught D&T, and the Year 8s are over halfway through their second year of it. So there is increasing 'feel' on the part of teachers for what D&T is, and last year's uncertainties are being replaced by growing certainty. But revelations about Key Stage 4 lurk just on the horizon, and promise a second wave of change. This article sets out to explore the change process that started in summer 1987 when 'technology' was identified as a subject to be included in the National Curriculum of all children between the ages of 5 and 16. It is worth first recalling briefly the formative events that resulted in the decision to implement a National Curriculum.

■ Enter the politicians

The National Curriculum was introduced to 'raise standards of attainment' which were perceived by the government to be generally too low. That perception was linked with a view that the nation's economic performance in a free-market, technological world was not as good as it should be. The often explicit but causally dubious connection is that improved 'standards' in education will improve the country's economic performance. The consultation document that announced the government's National Curriculum intentions (DES/WO, 1987) itself referred specifically to Callaghan's 1976 'Ruskin' speech that had initiated the 'great debate' about education.

The intention of the National Curriculum is to raise standards by:

- not allowing children to 'drop' subjects early;
- setting clear achievement objectives for children 'which the pupils themselves and their teachers, supported by parents and others, can work towards with confidence';
- ensuring that all children have access to 'broadly the same good and relevant curriculum' to ensure that 'the content and teaching of the various elements of the National Curriculum bring out their relevance to and links with pupils' own experiences and their practical applications and continuing value to adult and working life'. (DES/WO, 1987, para 8)

Each of these has implications for secondary school teachers of D&T. The first means that they have to accommodate and teach all pupils up to age 16 within the time allocated against other subjects. The second means developing the confidence effectively to teach new curricular requirements. The third implies an evaluation of what is now expected against what has previously been offered, raising issues of how what is new is better in these respects than what already exists.

Politicians have responded to a perceived need to raise educational standards by allocating considerable time and money to putting in place a bureaucracy to support the paperwork of a legally enforceable common curriculum. It is expected that teachers will work towards this curriculum with confidence. What is known about the way that teachers do their job and the way they respond to required change?

■ Enter the theorists

Educational theorists have attempted to describe and understand the processes of curriculum change. In the context of the wave of curriculum change in the USA in the early 1960s (which subsequently washed up on the shores of this country), Taba's book *Curriculum Development: Theory and Practice* (1962) argued for *institution-based* curriculum development, because of problems she perceived with both *top–down* and *localized* initiatives. Briefly explained, 'top–down' is an external requirement placed on teachers and having some weight of authority but not developed by the teachers, and 'localized' may be one person or department in a school which can be idiosyncratic and can result in much re-inventing of the wheel. In some ways, the National Curriculum is top–down (i.e. a centralized group has prescribed a curriculum that teachers have to implement). But, in fairness, the new curriculum requirement has insisted that practice should develop from what is already good and appropriate, and that teachers should thereby be able to exercise their professionalism. So there is clear potential at the level of each school for institution-based curriculum development within the National Curriculum approach – not least because the curricular requirement was not initially accompanied by resource material (by contrast with an earlier generation of courses such as Nuffield science). Taba's view anticipates that curriculum change requires both *changing the goals and means of institutions* and *changing individual teachers both cognitively and emotionally*. So for a change requirement as thoroughgoing as D&T in the National Curriculum, it may be expected that there will be considerable pressures on management practices at school level and on teachers' personal professionalism. This latter aspect indicates that work such as that of Hoyle (1980) will be applicable. Hoyle identified a difference between what he described as 'restricted' and 'extended' professionalism. A teacher operating in the former style tends to focus closely on the quality of the teacher–class interaction and not to worry too much about the educational world beyond – whereas the latter style takes a broader view of the educational context for the classroom interaction. The former style tends also to include valuing the building of relationships with pupils through long-term contact.

Chin and Benne (1969) recognized that different *strategies for effecting change* had within them different views of the nature of teachers. An *empirical–rational* strategy assumes rational, emotion-free and selfless ongoing evaluation by teachers that will cause them to identify and attempt to teach that which they can accept as being best for their pupils. A *normative–re-educative* strategy acknowledges that teachers operate within social frameworks, having roles and personal agendas and feelings. A *power–coercive* strategy implies that teachers are given no option but to change. Some of the rhetoric of the National Curriculum implies that teachers are being viewed as operating in a mode appropriate to the empirical–rational model – that they will inevitably recognize D&T as better than what has preceded it. The truth is that there is a strong element of the power–coercive, which pushes through the changed curricular *goals* but pays little attention to how teachers will be enabled to effect the required changes. It also ignores a reality of the normative–re-educative strategy: the need to recognize that teachers will need time and appropriate professional support in responding to change.

Out of the curriculum changes of the 1970s came such concepts as the 'teacher as researcher' (following Stenhouse, e.g. 1975) and the 'reflective practitioner' (following Schon, 1983). A subsequent generation of work on curriculum change has focused attention on the *careers* of individual teachers (e.g. Ball and Goodson, 1985). All of this points again to the centrality of the teacher in achieving effective curriculum change. Unfortunately, the politicians have tended to view teachers as being in need of direction and constraint, rather than attempting to understand the nature of their job and the support that they need in effecting change.

Also studied has been the significance of *subjects* (in the secondary sector – e.g. Goodson, 1988, chs. 8 and 10) which has demonstrated the powerful influence of the socially constructed school subject in the curriculum. A government that creates a newly amalgamated subject by committee has set up the conditions for considerable anguish on the part of those teachers whose professional support has come largely through familiarity with one of the previously existing, contributing subjects.

The reality is that all of these contributions to understanding how curriculum change is effected can be used to shed light both on what has happened and what needs to happen when change is required. Applying change theory to the establishment of D&T leaves questions about the expectations for implementation to which I shall return in the final section.

■ Enter the National Curriculum proposals

The July 1987 consultation document (DES/WO, 1987) identified Technology and Art–Music–Drama–Design as two of eight foundation subjects for all 15 to 16-year-olds, and made Home Economics and Business Studies optional. This caused wild throwing of hats in the air by those disparate groups who had been campaigning for many years for technology for all, but some discontented rumblings from groups who perceived a split between technology and design or the marginalization of design. Some Home Economics teachers began thinking of early retirement: others started an active lobbying campaign through their professional association (e.g. NATHE, 1988a, 1988b).

The interim report of the D&T working group (DES/WO, 1988) indicated that D&T was to draw on a range of secondary subjects and the final report (DES/WO, 1989) named them. Home Economics teachers found themselves included . . . but was it on the right terms when many CDT teachers, advisors and inspectors seemed to be trumpeting that CDT was doing it all anyway, and the others were welcome to make a small contribution? Some Business Studies and Information Technology teachers reflected that they had reasonable teaching curricula and adequate timetables anyway, and would have little time to contribute to D&T. Some Art departments looked forward to joining in – others preferred to wait and see what their own proposals would bring.

Some schools already had faculties that included some of the relevant subject specialists. In some instances this had been the result of LEA policy, or school policy, and had resulted in a relatively slow and careful bringing-together of specialists from a range of areas *that had been given time to work*. In some schools, D&T became an instrument that enabled sought-for cooperation. In others it became a crude thrusting-together of groups of subject specialists who had no previous thought of having to work together, but who had in place their own individual systems that enabled effective teaching and classroom survival.

■ Enter the contributing subjects

Some of the subject areas identified as contributing to D&T had experienced significant change in the years leading up to D&T implementation. But this change had been largely 'in subject'.

☐ CDT

The practical skills-based subjects of woodwork, metalwork and technical drawing were drawn into the CDT umbrella particularly when GCSE was introduced. Although 'the design approach' at that stage supplemented the 'skill development' approach, it was still possible for woodwork, metalwork and technical drawing teachers to identify themselves with particular GCSEs and to contribute from their expertise of particular materials and skills. They may not have liked having to use 'the design approach', although some took to it even when others did it reluctantly. Their main complaints about the new approach seem to have been that it detracted from the serious business of developing 'craftsmanship' by diverting time to non-craft activities and that it disadvantaged and disenchanted those children who enjoyed and achieved well in the practical work, but who didn't enjoy the 'academic' work. There was thus set up an interesting tension within a group of teachers who had often sought greater status for their subject, who had found their subject 'intellectualized' to an extent to give it that status, and who then found that some of the pupils who had previously been well-motivated became disenchanted. But the members of the new CDT departments often rubbed along quite well – the chippy, the tin-basher and the draughtsman (all male, of course) working in their own specialist rooms. Rumbling to each other at break times in the CDT staff cupboard usually defused any major disasters, and a new generation of CDT-wise Heads of Department and advisors jollied everyone along and moaned to each other that in some instances the labels on the doors had been changed but not much else. Although the required change of approach was quite significant, it didn't badly damage teacher confidence and competence because the materials, rooms and colleagues remained much the same, and any heart-searching was within the support group.

☐ HE

The advent of GCSE brought a set of syllabuses to Home Economics teachers which, again, built on traditional expertise and offerings. HE: Textiles and HE: Food were within the competence of teachers whose primary interest was in one of these areas and HE: Child Development offered the third area to complete the set. But, as with CDT, these new syllabuses required an 'investigative' approach, which led many HE teachers to change their style from the classical 'I demonstrate – you make' formula to one where pupils had to do far more selection and pre-analysis before they could actually get on and cook, or whatever. This change in emphasis (also brought in to enhance the subject's academic respectability) affected Child Development as well as the others, and one perceived result was the same as in CDT – that children who had previously been motivated to get on and enjoy an essentially practical subject were becoming disenchanted with 'all this writing and researching'. But, again, most teachers were able to take the required changes within their own competence and to find support from colleagues.

☐ BS

What had been essentially vocational courses in shorthand and typing found themselves having to respond to the remarkable pace of change in office practice in the world. Many BS teachers found themselves 'running to keep up' as typewriters, updated to electronic models, then became supplemented by networked computer terminals. They also often found themselves having to accommodate 'enterprise' activities within their teaching schemes. BS teachers tended to be rarer birds than their CDT and HE colleagues, and were more likely to have to turn to county-wide (rather than in-school) networks for support. But they have pressed on, and the ethos – often that of preparation for the vocational world of the secretarial office – has remained similar. In some schools, the teaching of Economics to A-level has been included in departmental responsibility, thus moving the department away from a vocational emphasis.

■ Enter the teachers and their pupils

Although research has been carried out on the day-to-day interactions between teachers and their pupils, this key educational interaction does not feature strongly in models of curriculum *change*. To my mind, that is a significant omission, because curriculum change strikes at the heart of a teacher's confident relationship with the pupils (the relationship implicit in but glossed over in the

National Curriculum proposals). Whatever view we have of the nature of the relationship between teacher and pupils, the essential requirement is that one person has a professional agenda that requires the assent of a group of pupils with their own agendas. To the usual requirements of classroom order, the D&T teacher adds safety concerns, pupils' individual practical work, personal practical skill, and a wide-ranging knowledge area that is never quite sure what the next practical or intellectual challenge will be.

Let me reflect on just one D&T lesson that I observed in the autumn term of 1990 when a class of pupils were carrying out food-based tasks in groups of about four. What enabled the teacher to run a successful lesson? In no special order, I would name years of experience, subject expertise (theoretical and practical), thorough preparation before the lesson, personality, familiarity with the room and its equipment and with the way children work, and (in this case) hours of work getting to grips with the new D&T Order and seeing how this particular lesson would contribute to the satisfaction of various statements within programmes of study and attainment targets.

There is little doubt that this teacher displayed confidence, although she told me that she *felt* less confident than usual because she was teaching something new – statutory D&T as opposed to familiar home economics. Confidence comes in part from working in a known context, and the National Curriculum was, then, a new and unknown context. Requiring teachers to teach and assess competently within what is in many ways a new given syllabus inevitably disturbs their confidence, especially when they feel that no-one 'out there' (or is it 'up there' in the case of a top–down development) has any more idea of what D&T is than they have themselves. Who are these 'others' who were supporting the teachers in this new requirement? Some LEA advisory teachers have worked hard to support their teachers, and some support material is now emerging nationally, but when critical early decisions were being made, the sense was very much of 'working out our own salvation'. And what happens when the supportive parents come along to parents' evening and ask 'Why isn't Jemima coming home each week with proper cakes like her Mum did?' – and the equivalent 'Why isn't Jimmy doing proper woodwork like his Dad did?'

I want to reflect further on what *actually* enables a teacher to perform with confidence and competence. The 'chemistry' between one teacher and one class is unique to that encounter. For the teacher, survival, confidence and competence are inextricably entwined. Knowing the aims for each lesson, knowing the subject thoroughly, knowing where things are, knowing how to trouble-shoot, are all essential aspects of successful teaching. For most D&T teachers, this means being thoroughly familiar with the work room and the jobs that the children are doing. It means having a system for dealing with tools and other equipment, work-in-progress, assessment, and generally keeping tabs on what the children are doing. It means having a 'feel' for the subject. D&T teachers have been that only since September 1990, having previously been CDT (and, before that, woodwork, metalwork or technical drawing) teachers or HE (and, before that, domestic science or needlework or cookery) teachers or BS (and, before that, shorthand and typing) teachers, and so on. All of them had a confidence and competence associated with particular ways of doing things needing particular familiarity with specialized equipment. And all were surviving with their own support structure.

One essential aspect of a teacher's support structure is the relationship with colleagues. Most of this is informal, and comes from just a few immediate colleagues. It is the break time or after-school conversation about a particular incident. It is having one or more colleagues to whom one can expostulate about the kids, the government, the Head, the caretaker, educationalists, whatever. And it is usually with people who share the same view of the nature of the job. So CDT teachers may well have been mutually supportive, as may have been HE teachers. And, because the nature of the job means that most break times and many lunch times are spent working with children, rather than in a staffroom, and because the departments may have been geographically remote, the support group tends to be small, and one of personal friendship.

■ So what has happened so far, and what does theory reveal?

A power–coercive strategy was used initially to bring into the legally enforceable curriculum of all children a 'subject' that many teachers welcomed. But such a power–coercive strategy takes no account of the realities of how teachers change. The government (mainly through the Curriculum Councils) now appears to attempt to operate an empirical–rational strategy, assuming that revealing what teachers have to do to teach D&T will result in them doing it. The more realistic normative–re-educative strategy would recognize the need to change teachers' values by giving them the time and means to re-examine their existing approaches. This conforms to an extent with the view of the teacher as a reflective practitioner, even though the initiative is from outside, rather than from the teacher.

Even if one assumes the more clinical, empirical–rational approach, one has to admit the possibility that some teachers will look at the new curriculum and will perceive aspects of it as less good than what is currently being taught. Such a conclusion also has to be permissible if one views the teacher as a reflective practitioner.

To put some flesh on the bones of these theories, let us consider the example of a comprehensive school. It is no more 'typical' or 'untypical' than all the others, but the observed responses resonate with many teachers that I talk with. The school had heads of art, BS, CDT and HE running healthy, but geographically and managerially separate, departments in the school year 1989–90. By healthy departments, I mean offering a range of courses appropriate to the pupils, producing above-average exam results, enjoying the approval of Head and parents, and including staff who feel valued in the school. The management structure of the school had tacitly supported a 'restricted' professionality by looking to individuals within small departments for successful teaching. There had been no formal structure of meetings or of information dissemination other than informal staff contact and a termly whole-staff meeting. However, when the National Curriculum turned out to be a real requirement, the Head instituted a structure of regular meetings of groups corresponding to National Curriculum 'subjects' and also a curriculum group of heads of these subjects.

The Head routinely sent any incoming information with 'technology' in its title to the head of CDT. Add to this the fact that the head of CDT was on a higher allowance than the others involved in D&T, one has in this school the initial feeling – reinforced at national and LEA level – that D&T will be mainly the responsibility of CDT. When the final report emerged, and the 1990–91 timetable had to be produced to include D&T in Year 7, initial management decisions had to be made. The Head suggested a confederation of existing subject departments to teach D&T. That confederation turned out to include BS, CDT and HE. Art decided not to be involved and the head of CDT was appointed as overseeing the working of the confederation. This immediately set up a different context for the response of teachers in the three departments – HE appearing to have to establish its role within D&T because there was no teaching to be done outside D&T; BS feeling 'tacked on' in a rather random way while still anticipating having its own teaching to do.

Professional concerns at this stage focused on how year groups were to be timetabled, what should be taught, how it should be taught, and how assessment should be carried out. It was felt that no relevant help was available from outside at this stage. Cases had to be argued in school about appropriate timetabling with a timetabler for whom D&T was just one small agenda item. Aspects of the implementation of this top–down curriculum imposition had to be fought from the bottom up in the school. *Personal* concerns focused on the change from working with a small group' of colleagues having shared understandings and frequent informal discussions, to a large group with formalized communication systems and uncertainty about understandings. For some, a concern was that a subject assembled by committee looked likely to have to be taught by committee, and the further erosion of the long-term teacher–class relationship was felt to be undesirable.

Although a meetings structure was set up, most of the people involved were not familiar with either the need to attend formal meetings or how to operate in this mode. For some, meetings were seen as a further imposition on valuable time that could be better spent on the 'real' tasks such as developing workschemes. Issues tended to appear to be divisive simply because they were seen from different subject perspectives. Rarely did the opportunity arise elsewhere for cross-subject conversation outside these formal meetings, not least because of the geographical separation of the departments and the ever-present time pressure. And there simply wasn't enough time to do everything when professionalism demanded that the need to develop material for the children to use came first. It is one thing to set up management structures in a school: it is quite another to enable teachers to work effectively within and through them.

Initial reactions differed across subjects. CDT operated from a relatively unthreatened position, feeling that the process and content were familiar. This relative confidence tended to assume that BS and HE were 'coming in' to the CDT fold. Some more traditional teachers saw the new proposals as yet another stage in watering down – a detrimental further moving away from skills towards 'designing' and less time again because of the drawing-in of BS. HE showed more uncertainty. There had to be more grappling with the Order to get the feeling that the current approach in HE teaching *was* in line with what was wanted. The concern was that only parts of HE would be represented (thus losing a coherent whole), and that the useful Child Development course would be lost. Thus there was concern about professional standing and a sense of being devalued and deskilled. BS tended to feel that their required involvement was the result of an arbitrary political decision (at national and school level) that found them having to contribute to teaching an unfamiliar subject to a new group of children, thus stretching their resources and bringing professional uncertainty. Responses to all of this had to be worked out in school. National and regional Inset tended to follow rather than lead.

The theory suggests that, if teachers are effectively to teach a new curriculum, they need time and professional support of various kinds. In testing my own observations against the experience of a range of teachers, my evidence thus far is that neither has been supplied in anything like the appropriate measure, and teachers are working out their own salvation at considerable personal cost. The remarkable thing is the tenacity of teachers in being determined that the children they teach shall not suffer a decline in quality of teaching. One cannot help wondering just how the National Curriculum in D&T is helping them to achieve this. An imposed top–down change needs to convince teachers of its value. When it appears to some to marginalize them, and to devalue their existing contribution, there is an uphill struggle. Ignoring the realities of subject-based support groups is a recipe for difficulty.

Teachers working in such circumstances are unlikely to accept a new package in its entirety. They will need to be convinced that what is now required is better than what they are currently doing. Their response tends quite reasonably to be incremental and discriminating. What actually happens is that a response evolves as particular problems are faced – e.g. how to arrange teaching groups to ensure curriculum coverage, how to assess, how to agree on interpretations. Conversations about these raise profound issues of professionalism that only gradually get addressed. Some features of the new curriculum will be found by some to enhance their teaching; others will hinder a particular style. What becomes clear to me is a determination to retain what works educationally – dare one say, to keep standards high – despite the pain caused by politicians who pay little heed to even the limited light shed by the theoreticians.

■ References

Ball, S.J. and Goodson, I.F. (eds) (1985) *Teachers' Lives and Careers*. Lewes: The Falmer Press.

Chin, R. and Benne, K.D. (1969) General strategies for effecting change in human systems. In W.G. Bennis, K.D. Benne and R. Chin (eds) *The Planning of Change* 2nd edn. New York: Holt, Rinehart and Winston.

Department of Education and Science and the Welsh Office (DES/WO) (1987) *The National Curriculum 5–16: A Consultation Document*. London: DES/WO.

Department of Education and Science and the Welsh Office (DES/WO) (1988) *National Curriculum Design and Technology Working Group Interim Report*. London: DES/WO.

Department of Education and Science and the Welsh Office (DES/WO) (1989) *Design and Technology for Ages 5 to 16*. London: DES/WO.

Goodson, I.F. (1988) *The Making of Curriculum: Collected Essays*. Lewes: The Falmer Press.

Hoyle, E. (1980) Professionalisation and deprofessionalisation in education. In E. Hoyle and J. Megarry (eds) *World Yearbook of Education 1980*. London: Kogan Page.

National Association of Teachers of Home Economics (NATHE) (1988a) The place of home economics in technology. Submission to the Design and Technology Working Group by NATHE. London: NATHE.

National Association of Teachers of Home Economics (NATHE) (1988b) *Technology in Home Economics*. London: NATHE.

Schon, D.A. (1983) *The Reflective Practitioner: How Professionals Think in Action*. London: Temple Smith.

Stenhouse, L. (1975) *An Introduction to Curriculum Research and Development*. London: Heinemann.

Taba, H. (1962) *Curriculum Development: Theory and Practice*. New York: Harcourt Brace Jovanovich, Inc.

5.7

Subject Subcultures and the Negotiation of Open Work: Conflict and Cooperation in Cross-curricular Coursework

C.F. Paechter

'Nobody in this school has got a complete oversight of what's happening environmentally, curriculum-wise. You know they turn to us two, but no, we're MEG environment, or NEA environment. We haven't got time to go and discover what's happening elsewhere, we're too busy throwing fuel on the fire, keeping those kids going.' (*Biology teacher, Lacemakers School*)

The research reported here is being carried out as part of the Cross-curricular Assessment Project, based at the Centre for Educational Studies, King's College, London, and supported by SEAC and the Training Agency. The Project was set up in September 1988 to develop cross-curricular projects which could be submitted as GCSE coursework, with any one piece of work being credited for two or more subjects. There are 28 schools or colleges in 11 LEAs involved in the work, with four of the schools currently being studied in more depth, so that issues arising from the negotiation and carrying out of cross-curricular coursework can be further clarified. The main research methods are: reflective participation, the researcher working with staff to develop plans for cross-curricular work, combined with (in the four case-study schools) interviews with the key participants, including students, at various points during the development of the work, and some observation of meetings, for example, with moderators.

In this article I want to consider the interaction between teachers' subject subcultures and open work, as it is manifested in relation to the planning and carrying out of cross-curricular work in fourth and fifth year classrooms. In doing this I shall raise some questions about the possibility of negotiating cross-curricular open coursework and tentatively suggest ways in which schools might proceed.

It should be stressed at the outset that I am using the two terms 'subject subcultures' and 'openness' in particular ways. Most importantly, I am treating the subject subculture as something that, however contested, is in certain contexts focused around an examination syllabus. The effects of assessment schemes on teachers' subcultural stances has not been taken sufficiently into account in the literature, although that may be because most of the work carried out pre-dates GCSE and the widespread appearance of substantial amounts of teacher-assessed coursework in Mode One (Examining Group-written) syllabuses. With the introduction of National Curriculum assessment, it is increasingly clear that the interaction between subject subculture and teacher assessment will need further study, given the growth in importance of cross-curricular links and the influence of schemes of assessment in all areas of school life.

■ Openness in the context of assessment

An open task is one that leaves some of the decisions to be made regarding the work within, or outcomes of, that task up to the learner, in negotiation with the teacher. However, within the context of coursework being carried out for assessment in a public examination, the notion of openness has to be modified into one of openness within external constraints. While in the normal classroom situation a variety of constraints affect the work carried out, these are likely to be concerned either with such things as available equipment, or with intended learning outcomes, and are in this sense internal to the situation. The additional constraints imposed when work is to be submitted for assessment, on the other hand, may not be concerned with learning at all, but rather with summative assessment and with 'assessability', whether a piece of work is sufficiently aligned with assessment criteria for it to be possible to use those criteria for its assessment. The emphasis in planning and carrying out a task moves from the process involved to the end points to be reached (where such end-points may include having involved a particular set of processes). The ever-present spectre of assessment means that the teacher has always to consider whether the work being carried out, however intrinsically valuable, will also satisfy the assessment scheme.

■ Subject subcultures and the concept of 'assessability'

One feature of many subject subcultures is their tendency to develop within teachers a unified or at least dominant view of the nature of their particular subject. The literature in this area rightly stresses that school subject teachers do not form homogeneous subject groups, but that there is debate and contestation between segments within an overall subject group (St-John Brooks, 1983; Goodson, 1985; Cooper, 1984, 1985a, 1985b; Ball and Lacey, 1980; Ball, 1985). However, the examination syllabus and assessment scheme, superficially at least, provide a unifying focus for any subcultural grouping, because they lay down a view of what product will result from teachers' classroom work, whatever those teachers' personal stances regarding the nature of their subject. The need to assess means that the teacher has to internalize a view of what counts as, for example, a 'good piece of mathematics', which is particular to the assessment scheme being used. Whatever the teacher's personally developed views about what counts as 'good mathematics', within the context of assessed coursework what matters is the definition implicit in the assessment scheme.

While there is limited freedom of syllabus choice at departmental level, so that a united department is usually able to select a syllabus that reflects at least a broad consensus, teachers as individuals and groups have to adapt themselves to the view of their subject exemplified in the syllabus chosen. This is not always clear-cut, for syllabuses and other curriculum documents are open to interpretation (Ball and Bowe, 1990; Scott, 1990), and the teacher has to be aware not only of the departmental view of an assessment scheme, but also of the view of the moderator, who may see it differently. In negotiating coursework with students, teachers therefore have to constantly make decisions about the assessability of a piece of work under a particular assessment scheme. A number of factors have to be considered: will the work proposed take up approximately the length of time recommended by the syllabus for spending on coursework; will it require a lot of oral assessment with little documentary evidence to show to the moderator; is there enough obvious content involved to satisfy content requirements for higher grades; will it demand so much teacher input that the student ends up being downgraded for requesting help (a problem with some science assessment schemes); or is it likely to be just plain difficult to mark?

It seems to me, therefore, that whatever an individual teacher's pedagogic views, he or she must develop and adopt an internalized subcultural stance that is aligned with the requirements of the assessment scheme being used. The assessment criteria have to, in a sense, enter the teacher's 'thinking as usual' (Schutz, 1964) so that they are, at least on a day-to-day level, uncontested.[1] Internalizing syllabus requirements in this way will involve different degrees of compromise for different people, but remains necessary for all teachers because of the need to make rapid decisions about assessability during teacher–student negotiation. Thus a teacher working on examination coursework will be dealing simultaneously both with a view of the subject expressed through pedagogical aims, and the internalized, but possibly alternative, subject-based requirements of assessment for a particular syllabus.

Internalized notions of assessability are developed from the teacher's reading of the constraints of the assessment scheme. Such schemes may themselves contain implicit assumptions about what sort of work is assessable, as well as about the nature of assessment. For example, most science assessment schemes are atomistic, giving marks for skills or groups of skills; this is not the case in most other subjects. In the humanities it is generally assumed that the work to be handed in will consist of written accounts, although this is not explicitly stated, and Examining Groups have indicated their willingness to accept work in other forms, such as annotated photographs, if it still indicates that the student has grasped the required concepts.

Sometimes the requirements of the assessment scheme itself make extra administrative demands on the teacher which lead to a perceived need to limit the possible outcomes of a piece of work. For example, one mathematics teacher (under pressure from senior management to do more 'adventurous' work) felt that the need to draw up a marking grid for each task set meant that he had to limit the choice available to the students:

> 'It's dead easy to sit in your office and say oh . . . why can't you do this, why can't you do that . . . It's something else to get in there and work with the children and then assess it afterwards which is the hardest job there is, you know that. I mean, I've got to go through each of these four tasks and align everything on to there [SMP marking grid, which has to be specifically adapted for each task]. I don't want to do that for these four tasks and do it again in March for the next four.' (*Mathematics teacher, Lacemakers School*)

The need to prioritize assessability when negotiating GCSE coursework imposes extra pressures on teachers' pedagogic decisions. The trade-off between subject outcomes and more general learning may become more critical, as it is less acceptable for the teacher to say that while the students did not learn a lot of, say, history, they did learn other things (such as independence). If a task does not result in anticipated outcomes, not only is subject time 'lost' (because teachers find it difficult to integrate coursework with normal class activities) but examination results may be affected. If students feel that their valuable GCSE time has been 'wasted' by the teacher, teacher–student relationships may be imperiled.

In consequence, when negotiating examination coursework, the teacher will tend to impose implicit or explicit constraints, arising out of the assessment scheme, on the students and their work. These will clearly affect discussions between teacher and student, particularly as they may seem arbitrary to the student (or, indeed, the teacher). The discontinuity between the requirements of the assessment scheme and a particular teacher's normal teaching style may be one reason why in many cases coursework does not emerge seamlessly out of normal classwork, as was the original intention of the GCSE. This is especially likely to be the case where there is a conflict between a teacher's subcultural stance regarding pedagogy and the syllabus-based stance regarding assessment, leading assessed work to be seen as somehow different from work focused on learning. Particularly where a teacher feels that the assessment scheme being used imposes tight constraints on what is acceptable, he or she may feel the need to 'signal' to the students that they are being assessed. Many teachers, for example, give students copies of their assessment schemes, in an attempt to ensure that they meet relevant criteria.

■ Assessment-focused subcultures and open work

When negotiating open work in the secondary classroom, a teacher would normally make use of his or her subject subcultural background. This makes it possible to make rapid decisions about a piece of work; whether it is likely to produce outcomes relevant to the subject area; whether it will stretch the student; which knowledge areas are likely to be involved. Because of the nature of the work, open tasks are likely to be characterized by frequent but often quite brief negotiations between teacher and students, with the teacher offering advice, some direction and help where necessary. In most cases this advice will be mediated by the teacher's aims and beliefs about the subject and the role of open work within it; the internalized nature of these allows judgements and decisions to be made rapidly, with little need for prolonged reflection.

Within the context of assessed coursework, the teacher's assessability-related subcultural background is also brought into play. As was stated earlier, an initial aim of the Cross-curricular Assessment Project was to work with schools to develop cross-curricular tasks that could be assessed for more than one subject in the GCSE examination. Although it was not explicitly stated, there was an assumption that a large proportion of these would be open tasks of the kind developed in earlier technology-related projects. This has not in fact occurred, and indeed there has been a tendency over time for projects to become less, rather than more, open. I believe that this is at least in part a result of the need for teachers to depend on internalized subcultural norms when negotiating work with students in open situations.

Cross-curricular work in the Project schools has taken a number of forms: suspension of the timetable for all or part of a year group; subject-based modules that link to or contribute to additional subjects; team-taught groups; fieldwork-based projects; work from one subject being developed into another; links between individual teachers for all or some of their coursework requirements; individually negotiated projects. Most of these models involve a teacher spending some of his or her time developing work with students that will be examined in a subject not that teacher's own. In this situation, the teacher does not have the requisite subcultural knowledge necessary for rapid negotiation with students, and it is here that plans for open work can break down. Furthermore, the subcultural norms, particularly regarding methods of assessment, may not be the same in all subjects, and in some cases can conflict.

Our work suggests that teachers have very little idea of what goes on in other subjects. Even where there has been extensive curriculum audit, or where teachers have read each other's syllabuses, there is little understanding of what 'covering' a particular topic in another subject really means. For example, it is often believed that the carrying out of mathematical operations within a piece of work will make it acceptable as a piece of mathematics coursework, even when this has been achieved by the use of worked examples provided by the teacher; the requirement for the student to have some understanding of the processes involved is not appreciated. Teachers are also relatively unaware of the standards required in other subjects, and can often demand work of too high a standard or accept that which is not good enough. Two teachers, one from Art and the other from English, both remark on this in connection with a project in which students write and illustrate children's books, which was initially carried out independently but is now a piece of joint work:

> 'I mean, one of the problems that was coming up for the first time in the English department . . . they had books that they were very pleased with, because obviously they were looking at them from a literature point of view, and when they actually flashed them at us for an artistic point of view, they were a bit disappointed because we didn't seem thrilled to bits. You know, we have similar problems the other way. We have shown them books that we thought the graphics were great on and they have read the stories and put their head in their hands and said, "I don't believe this is happening".' (*Head of Art, Shipbuilders School*)

> 'Originally, English teachers were sort of asking them to illustrate *everything*, which is ridiculous, whereas in fact, you know, Sally, the head of Art was saying, you're not really, you're talking about one or two things . . . so we as English teachers were actually asking something very unrealistic of our students, and in terms of GCSE, you know, one of the original criticisms was overload for the students. It was our ignorance as English teachers, it was very well intended, but we didn't realize what we were asking the kids to do. Now the fact that we've got this tie-up for teachers, and the fact that we've actually got down to discuss it, what we were asking them to do, has made it easier for the kids.' (*Head of English, Shipbuilders School*)

Working in a cross-curricular way can challenge subculturally based assumptions and lead to unease. This mathematics teacher has recently started supporting, for one lesson a week, a 12-week Technology module, taught mainly by a CDT teacher. The work produced will become a piece of mathematics coursework, so he has in effect had to relinquish the responsibility for the negotiation and production of mathematics coursework to a teacher of another subject:

> '. . . he's much more used to long projects and things taking a long time and I've never been involved in a piece of work that's taken 4 weeks before, really. I mean if we set one of these in a Maths lesson it'll have to be done in 4 weeks, and then we say finito, signed, sealed, delivered, here you go, and he's not worried if they're only on

their 2nd drawing or whatever, and I'm a bit more . . . I want to see it getting somewhere – I'd rather see it coming together quicker than it is – I mean I hope it'll come together in the end.' (*Mathematics teacher, Lacemakers School*)

A teacher negotiating open coursework within one subject area has as a main concern the assessability of that coursework, and in particular the boundaries of subject matter, approach and teacher input implicit in marking schemes. These can be difficult to convey to teachers of other subjects, however well-intentioned, because words have subtly different shades of meaning between one subject and another, and what counts as unacceptable help in one subject may be seen as normal practice in another. For example, many teachers see the practice of re-drafting in English as a form of cheating, of unfair enhancement of students' work, whereas it is considered by English teachers and examiners as normal good practice. English teachers, on the other hand, are apt to consider students' work using quotation from reference sources as 'not their own work' and therefore unacceptable. The difficulties that can occur when teachers of one subject intervene in work that is to be used as coursework in another are illustrated by the experiences of two teachers, one science, one geography, with students taking measurements on a field trip:

'If I am going to compare . . . shall we say, discharges at a particular site . . . I need to know that there is consistency of measurement there. Science on the other hand maybe looking for how can you . . . having measured at one site . . . one location . . . how can we improve it? . . . Now that's fine as long as they can then ensure that the information that I want collected is collected.' (*Head of Geography, Stitchers School*

'The design [assessment category] was OK, because I did a lot before we went away . . . But again what happened was that in order to get the maximum 6 [marks], they need to get across some sort of problem and change their design. Now what happened was when they came across the problems, and, you know, various different things they needed to do, they couldn't quite get it. Instead of allowing them to sit down and redesign and evaluate what they had done, a member of staff would plough straight in and say, don't do that, do it like this. It's understandable because Geography need the results and they're not interested in the way which they're got, but it detracts from the science.' (*Science teacher, Stitchers School*)

Teachers are acutely aware of the gap between others' perceptions of their subject and their own, and in particular of the differences between subjects regarding what is acceptable for assessment. Consequently, when cross-curricular projects are being conceived, there is an enormous temptation to 'tie everything down' in such a way that no-one can misunderstand. This seems particularly to be the case with science teachers, who are used to an atomistic mode of assessment, and who find it difficult to work with those that take a more holistic view. This tendency to get everything into what one teacher called 'tidy-boxes' works directly against the development of open work, as teachers attempt to control all the possible outcomes so that their colleagues are unable to let them down, or vice versa.

■ Some suggestions

The difficulties for the planning and carrying out of tasks involving multiplely assessed open coursework tasks arise from the overlap of complexities in all the different components. The negotiation of open work is in itself complex and relies to a large extent on ongoing, subculturally rooted judgements on the part of the teacher. If such work is to be assessed, a further subcultural layer is introduced, with the teacher having to bear in mind not only pedagogic considerations but also notions of assessability implicit in an interpretation of the criteria for assessment. If the intention of dual accreditation is introduced to this already complex scene, requiring teachers to work within the subcultural norms underlying the assessment of subjects other than their own, the whole project can collapse.

At this point the tendency has been for teachers to move away from open work towards coursework tasks that are much more constrained. However, this is not necessarily the best way forward. A partial solution being developed in some schools is the abandonment of dual accreditation while still carrying out joint tasks. One area where this has been successful is in work done in or around CDT subjects. Most CDT courses require a long individual project, and schools have found that the length of this task and its position late in the course makes it difficult to use as the basis of joint coursework. However, it is unusual for students not to have carried out other similar but shorter projects earlier in the course, and these can be set up in such a way that they provide or contribute to coursework tasks for another subject. In one school a complete piece of mathematics coursework is produced in this way. In another the work done in the fourth year in Design and Realization is developed and marked in Business Studies in the fifth year. Some schools are also making use of areas that are not normally assessed, developing Integrated Humanities work from PSE, for example, although again this can run into problems if those teaching PSE are not sufficiently aware of the requirements of humanities assessment.

Timetable manipulations can also be a way to ensure that all relevant subcultural assumptions are catered for. These can take a number of forms and some are more successful than others. Timetable suspension, which has been tried in several schools, can in fact exacerbate the problems, as students tend to end up working with only one or two teachers throughout, and have less access than usual to teachers of the other subjects involved. The tendency in this kind of project for teachers to take students off-site and to work through the lunch hour means that staff are also less accessible to each other and so tend to carry on without consultation once the project has begun. The only case among the Project schools in which timetable suspension has resulted in dual-accredited coursework is one in which the students were away on a field trip accompanied by teachers from all the relevant subjects.

Team-teaching has been more successful, both where one teacher has major responsibility for the group and is occasionally supported by the other, and where two teachers work together with a double-size group. In the technology module referred to earlier, it was noticed that students tended to save mathematical queries for the days when the mathematics teacher was present, and were encouraged to do so by the CDT specialist who did most of the teaching. This system also ensured that the students recorded their work in such a way that it could be assessed relatively easily for mathematics. Where it has not been possible to put two teachers in one classroom, even once a week, time has sometimes been given to the teachers involved in a project to meet on a regular basis while it is going on so that they are able to share ideas and pick up any problems arising at an early stage. We have found that where joint coursework has broken down it has tended to be in cases where, while there was initial discussion, this was not continued while the work was being carried out. Close and continued communication, however informal, is essential. This is the case even with closed cross-curricular tasks, but is especially so with regard to open work.

If openness is to be preserved in cross-curricular coursework, teachers need time to discuss in detail the nature of their various subjects, as well as the particular requirements of the planned task, in terms of both pedagogy and assessability. In this way, it may be possible to convey to others key points at which it is important to ensure that advice is obtained from the subject specialist. In coming to this process, teachers will need to abandon their preconceptions of others' subjects, particularly those bound up with subject status, so that the central issues for all concerned can be made clear. This is not as easy as one might at first imagine, and in one school such discussion appears to have strengthened prejudices held by heads of high-status departments, concerning the 'superior rigour' of their subjects. A little knowledge can indeed be a dangerous thing, and it is essential that teachers come to this process in a spirit of respect for the mores of each others' subjects. One teacher described in this way the essential quality that he saw as necessary for good co-operation:

'You have got to have people who are generous . . . you can't legislate for that generosity, you know, you can't actually put it down on the application form and say, "you must be a generous person", but that is what is needed, and if it isn't there it does become difficult.' (*Art teacher, Lacemakers School*)

Acknowledgement

The research reported in this paper was carried out as part of a project entitled 'Cross-curricular Assessment Through Coursework for GCSE' and funded by the

Schools Examinations and Assessment Council and the TVEI Unit of the Department of Employment.

Note

(1)   The experience of the Project is that this internalization is quite deep, many teachers, for example, finding it hard to think outside a mind-set developed in the early years of the GCSE. There is an observed tendency for teachers to continue to adhere to regulations that have since been lifted, and the cautious interpretations of coursework assessment schemes developed at that time are hard to challenge.

■ References

Ball, S.J. and Lacey, C. (1980) Subject disciplines as the opportunity for group action: a measured critique of subject subcultures. In P. Woods (ed.) *Teacher Strategies*, pp. 149–77. London: Croom Helm.

Ball, S.J. (1985) English for the English since 1906. In I. Goodson (ed.) *Social Histories of the Secondary Curriculum*, pp. 53–88. Lewes: Falmer Press.

Ball, S.J. and Bowe, R. (1990) *Subject to Change? Subject Departments and the Implementation of National Curriculum Policy: An Overview of the Issues.* Paper presented to the British Educational Research Association Annual Conference, September.

Cooper, B. (1984) On explaining change in school subjects. In I.F. Goodson and S.J. Ball *Defining the Curriculum: Histories and Ethnographies*, pp. 45–63. Lewes: Falmer Press.

Cooper, B. (1985a) *Renegotiating Secondary School Mathematics.* Lewes: Falmer Press.

Cooper, B. (1985b) Secondary school mathematics since 1950: reconstructing differentiation. In I. Goodson (ed.) *Social Histories of the Secondary Curriculum*, pp. 89–119. Lewes: Falmer Press.

Goodson, I. (1985) Subjects for study. In I. Goodson (ed.) *Social Histories of the Secondary Curriculum*, pp. 343–67. Lewes: Falmer Press.

Scott, D. (1990) *Coursework and Coursework Assessment in the GCSE.* Cedar Report 6, University of Warwick.

St-John Brooks, C. (1983) English: a curriculum for personal development? In M. Hammersley and A. Hargreaves (eds) *Curriculum Practice: Some Sociological Case Studies*, pp. 37–59. Lewes: Falmer Press.

Schutz, A. (1964) The Stranger. In B.R. Cosin, I.R. Dale, G.M. Esland, D. MacKinnon and D.F. Swift (eds) (1971) *School and Society*, pp. 27–33. London: Routledge and Kegan Paul.

Index